D1218716

Fire Safety Aspects of Polymeric Materials

VOLUME 5 ELEMENTS OF POLYMER FIRE SAFETY AND GUIDE TO THE DESIGNER

Report of

The Committee on Fire Safety
Aspects of Polymeric Materials

NATIONAL MATERIALS ADVISORY BOARD
Commission on Sociotechnical Systems
National Research Council

Publication NMAB 318–5
National Academy of Sciences
Washington, D.C.
1979

VOLUME 5 – ELEMENTS OF POLYMER FIRE SAFETY AND GUIDE TO THE DESIGNER

Fire Safety Aspects of Polymeric Materials

265 Post Road West, Westport, CT 06880

Printed in U.S.A.
Library of Congress Card No. 77-79218
I.S.B.N. 0-87762-226-4

NOTICE

The project that is the subject of this report was approved by the Governing Board of the National Research Council, whose members are drawn from the councils of the National Academy of Sciences, the National Academy of Engineering, and the Institute of Medicine. The members of the committee responsible for the report were chosen for their special competences and with regard for appropriate balance.

This report has been reviewed by a group other than the authors according to procedures approved by a Report Review Committee consisting of members of the National Academy of Sciences, the National Academy of Engineering, and the Institute of Medicine.

This study by the National Materials Advisory Board was conducted under Contract No. 4-35856 with the National Bureau of Standards.

Printed in the United States of America.

FOREWORD

This volume is one of a series of reports on the fire safety aspects of polymeric materials. The work reported here represents the results of the first in-depth study of this important subject. The investigation was carried out by a committee of distinguished polymer and fire technology scholars appointed by the National Academy of Sciences and operating under the aegis of the National Materials Advisory Board, a unit of the Commission on Sociotechnical Systems of the National Research Council.

Polymers are a large class of materials, most new members of which are man-made. While their versatility is demonstrated daily by their rapidly burgeoning use, there is still much that is not known or not widely understood about their properties. In particular, the burning characteristics of polymers are only now being fully appreciated and the present study is a landmark in the understanding of the fire safety of these ubiquitous materials.

In the first volumes of this series the committee has identified the limits of man's knowledge of the combustibility of the growing number of polymeric materials used commercially, the nature of the by-products of that combustion, and how fire behavior in these systems may be measured and predicted. The later volumes deal with the specific applications of polymeric materials, and in all cases the committee has put forth useful recommendations as to the direction for future actions to make the use of these materials safer for society.

Harvey Books, Chairman
Commission on Sociotechnical Systems

ABSTRACT

This is the fifth volume in a series. The fire safety aspects of polymeric materials are examined with primary emphasis on human survival. This volume is a summary of the first four volumes of the series, which deal with materials: state of the art; test methods, specifications, and standards; smoke and toxicity; and fire dynamics and scenarios. Later volumes deal with applications to aircraft, buildings, land vehicles, ships, and mines and bunkers.

The purpose of this volume is to provide an outline of the basis or disciplinary information relating to fire safety of polymeric materials for designers and others concerned with the problem, such as fire chiefs, fire inspectors, and fire investigators. Readers wishing additional detail are referred to the appropriate basic volumes.

VOLUMES OF THIS SERIES

Volume 1	Materials: State of the Art
Volume 2	Test Methods, Specifications, and Standards
Volume 3	Smoke and Toxicity
Volume 4	Fire Dynamics and Scenarios
Volume 5	Elements of Polymer Fire Safety and Guide to the Designer
Volume 6	Aircraft: Civil and Military
Volume 7	Buildings
Volume 8	Land Transportation Vehicles
Volume 9	Ships
Volume 10	Mines and Bunkers

PREFACE

The National Materials Advisory Board (NMAB) of the Commission on Sociotechnical Systems, National Research Council, was asked by the Department of Defense Office of Research and Engineering and the National Aeronautics and Space Administration to "initiate a broad survey of fire-suppressant polymeric materials for use in aeronautical and space vehicles, to identify needs and opportunities, assess the state of the art in fire-retardant polymers (including available materials, production, costs, data requirements, methods of test and toxicity problems), and describe a comprehensive program of research and development needed to update the technology and accelerate application where advantages will accrue in performance and economy."

In accordance with its usual practice, the NMAB convened representatives of the requesting agencies and other agencies known to be working in the field to determine how, in the national interest, the project might best be undertaken. It was quickly learned that wide duplication of interest exists. At the request of the other agencies, sponsorship was made available to all government departments and agencies with an interest in fire safety. Concurrently, the scope of the project was broadened to take account of the needs enunciated by the new sponsors as well as those of the original sponsors.

The total list of sponsors of this study now comprises Department of Agriculture, Department of Commerce (National Bureau of Standards), Department of Interior (Division of Mine Safety), Department of Housing and Urban Development, Department of Health, Education and Welfare (National Institute for Occupational Safety and Health), Department of Transportation (Federal Aviation Administration, Coast Guard), Energy Research and Development Administration, Consumer Product Safety Commission, Environmental Protection Agency, and Postal Service, as well as the original Department of Defense and National Aeronautics and Space Administration.

The Committee was originally constituted on November 30, 1972. The membership was expanded to its present status on July 26, 1973. The new scope was established after presentation of reports by liaison representatives covering needs, views of problem areas, current activities, future plans, and relevant resource materials. Tutorial presentations were made at meetings held in the Academy and during site visits, when the Committee or its panels met with experts and organizations concerned with fire safety aspects of polymeric materials. These site visits (upwards of a dozen) were an important feature of the Committee's search for authentic information. Additional inputs on foreign fire technology were supplied by the U.S. Army Foreign Science and Technology Center and NMAB Staff.

This study in its various aspects is addressed to those who formulate policy and allocate resources. A sufficient data base and bibliography has been supplied to indicate the breadth of the study.

ACKNOWLEDGMENTS

In the preparation of this volume, the committee is greatly indebted to its member, Mr. Donald G. Smillie, who prepared the first working draft for its consideration.

I acknowledge with gratitude the assistance in this project of Dr. Robert S. Shane, NMAB Consultant, and Miss Carolyn A Tuchis, his able secretary.

Dr. Herman F. Mark, Chairman

NATIONAL MATERIALS ADVISORY BOARD

Dr. John R. Hutchins III
Vice President and Director of
Research and Development
Technical Staff Division
Corning Glass Works
Sullivan Park
Corning, NY 14830

Dr. James R. Johnson
Executive Scientists and Director
Advances Research Program Laboratory
3M Company
P. O. Box 33221
St. Paul, MN 55133

Mr. William D. Manly
Senior Vice President
Cabot Corporation
125 High Street
Boston, MA 02110

Dr. James W. Mar
Professor, Aeronautics and Astronautics
Building 33-307
Massachusetts Institute of Technology
Cambridge, MA 02139

Dr. Frederick T. Moore
Industrial Advisor
Industrial Development and
Finance Department
World Bank
1818 H Street, N.W., Room D422
Washington, DC 20431

Dr. Nathan E. Promisel
Consultant
12519 Davan Drive
Silver Spring, MD 20904

Dr. Allen S. Russell
Vice President-Science and Technology
Aluminum Company of America
1501 Alcoa Building
Pittsburgh, PA 15219

Dr. Jason M. Salsbury
Director Chemical Research Division
American Cyanamid Company
Berdan Avenue
Wayne, NJ 07470

Dr. John J. Schanz, Jr.
Assistant Director, Center for
Policy Research
Resources for the Future
1755 Massachusetts Avenue, N.W.
Washington, DC 20036

Dr. Arnold J. Silverman
Professor, Department of Geology
University of Montana
Missoula, MT 59801

Dr. William M. Spurgeon
Director, Manufacturing and
Quality Control
Bendix Corporation
24799 Edgemont Road
Southfield, MI 48075

Dr. Morris A. Steinberg
Director, Technology Applications
Lockheed Aircraft Corporation
Burbank, CA 91520

Dr. Roger A. Strehlow
Professor, Aeronautical and
Astronautical Engineering
University of Illinois at Urbana
101 Transportation Building
Urbana, IL 61801

Dr. John E. Tilton
Professor, Department of Mineral
Economics
221 Walker Building
Pennsylvania State University
University Park, PA 16801

NMAB STAFF:

W. R. Prindle, Executive Director
R. V. Hemm, Executive Secretary

NATIONAL MATERIALS ADVISORY BOARD
COMMITTEE ON FIRE SAFETY ASPECTS OF POLYMERIC MATERIALS

Chairman

HERMAN F. MARK, Emeritus Dean, Polytechnic Institute of New York, Brooklyn.

Members

ERIC BAER, Head, Division of Macromolecular Science, Case-Western Reserve University, Cleveland, Ohio.

JOHN M. BUTLER, Manager, Organic and Polymer Synthesis, Monsanto Research Corp., Dayton, Ohio

RAYMOND FRIEDMAN, Vice President and Scientific Director, Factory Mutual Research Corp., Norwood, Massachusetts

ALFRED R. GILBERT, Manager, Organic and Polymer Chemistry Branch, Chemistry Laboratory, Corporate R&D, General Electric Company, Schenectady, New York.

CARLOS J. HILADO, Research Professor, Institute of Chemical Biology, University cf San Francisco, California

RAYMOND R. HINDERSINN, Section Manager, Research Center, Hooker Chemical Co., Niagara Falls, New York

WILLIAM C. HUSHING, Vice President, Kensington Management Consultants, Bath, Maine

FRANK E. KARASZ, Professor and Co-Director, Materials Research Laboratory, Polymer Science & Engineering Dept., University of Massachusetts, Amherst

TED A. LOOMIS, Professor and Chairman, Department of Pharmacology, School of Medicine, University of Washington, Seattle

RICHARD S. MAGEE, Professor, Department of Mechanical Engineering, Stevens Institute of Technology, Hoboken, New Jersey

ELI M. PEARCE, Head, Department of Chemistry, Professor of Polymeric Chemistry and Chemical Engineering, Polytechnic Institute of New York, Brooklyn

CHARLES F. REINHARDT, Director, Haskell Laboratory for Toxicology and Industrial Medicine, E. I. duPont Nemours and Co., Newark, Delaware

ARNOLD ROSENTHAL, Celanese Research Company, Summit, New Jersey

DONALD G. SMILLIE, Chief Structures Development Engineer, McDonnell Douglas Corp., Long Beach, California

GIULIANA C. TESORO, Adjunct Professor, Fibers and Polymers Laboratories, Massachusetts Institute of Technology, 278 Clinton Avenue, Dobbs Ferry, New York

NICHOLAS W. TSCHOEGL, Professor, Department of Chemical Engineering, Division of Chemistry and Chemical Engineering, California Institute of Technology, Pasadena

ROBERT B. WILLIAMSON, Associate Professor of Engineering Science, Dept. of Civil Engineering, University of California, Berkeley

Liaison Representatives

JEROME PERSH, Staff Specialist for Materials and Structures, ODDRE, Department of Defense, Washington, D.C.

ERNEST B. ALTERKRUSE, Chief, Department of Clinics, Moncrief Army Hospital, Fort Jackson, South Carolina

ALLAN J. McQUADE, U.S. Army Natick Laboratories, Natick, Massachusetts

GEORGE R. THOMAS, Department of the Army, Army Materials and Mechanics Research Center, Watertown, Massachusetts

HERBERT C. LAMB, Naval Facilities Engineering Command, Alexandria, Virginia

DANIEL PRATT, Naval Ships Engineering Center, Prince George's Center, Hyattsville, Maryland

GEORGE SORKIN, Naval Sea Systems Command, Washington, D.C.

JACK ROSS, Wright-Patterson Air Force Base, Ohio

BERNARD ACHHAMMER, National Aeronautics and Space Administration, Advanced Research and Technology Division, Washington, D.C.s,

JOHN A. PARKER, National Aeronautics and Space Administration, Ames Research Center, Moffett Field, California

ARNOLD WEINTRAUB, U.S. Department of Energy, Washington, D.C.

DAVID FORSHEY, Department of the Interior, Bureau of Mines, Washington, D.C.

CLAYTON HUGGETT, Center for Fire Research, National Bureau of Standards, Washington, D.C.

LESLIE H. BREDEN, Fire Safety Engineering Division, Center for Fire Research, National Bureau of Standards, Washington, D.C.

IRVING LITANT, Department of Transportation (DoT), Systems Center, Kendal Square, Cambridge, Massachusetts

GEORGE BATES, JR., Aircraft Division, DoT, Federal Aviation Administration, National Aviation Facilities Experimental Station, Atlantic City, New Jersey

RALPH RUSSELL, Aircraft Division, DoT, Federal Aviation Administration, National Aviation Facilities Experimental Station, Atlantic City, New Jersey

PAUL W. SMITH, Aviation Toxicology Laboratory, Civil Aeromedical Institute, Oklahoma City

ROBERT C. McGUIRE, DoT, Federal Aviation Administration, Washington, D.C.

THOMAS G. HOREFF, DoT, Federal Aviation Administration, Washington, D.C.

DANIEL F. SHEEHAN, DoT, U.S. Coast Guard, Washington, D.C.

WILLIAM J. WERNER, Department of Housing and Urban Development, Washington, D.C.

DONALD L. MOORE, Department of Housing and Urban Development, Washington, D.C.

IRVING GRUNTFEST, Environmental Protection Agency, Washington, D.C.

NELSON GETCHELL, U.S. Dept. of Agriculture, Beltsville, Maryland

WILSON A. REEVES, Cotton Finishes Laboratory, U.S. Dept. of Agriculture, New Orleans, Louisiana

HERBERT W. EICHNER, Forest Service, U.S. Dept. of Agriculture, Forest Products Laboratory, Madison, Wisconsin

RICHARD E. WIBERG, National Institute for Occupational Safety and Health, Rockville, Maryland

JAMES RYAN, Consumer Product Safety Commission, Bethesda, Maryland

JURGEN KRUSE, Materials Research Div., Office of Postal Science and Technology, Rockville, Maryland

Technical Advisors:

IRVING N. EINHORN, Division of Materials Science and Engineering, Flammability Research Center, College of Engineering, University of Utah, Salt Lake City

MAUREE W. AYTON, Science Information Analyst, Congressional Research Service, Library of Congress, Washington, D.C.

National Materials Advisory Board Staff:

ROBERT S. SHANE, National Materials Advisory Board, National Academy of Sciences, Washington, D.C.

CONTENTS

CHAPTER 1

INTRODUCTION

1.1 Scope and Methodology of the Study

The charge to the NMAB Committee on Fire Safety Aspects of Polymeric Materials was set forth in presentations made by the various sponsoring agencies. Early in its deliberations, however, the committee concluded that its original charge required some modification and expansion if the crucial issues were to be fully examined and the needs of the sponsoring organizations filled. Accordingly, it was agreed that the committee would direct its attention to the behavior of polymeric materials in a fire situation with special emphasis on human-safety considerations. Excluded from consideration were firefighting, therapy after fire-caused injury, and mechanical aspects of design not related to fire safety.

The work of the committee includes (1) a survey of the state of pertinent knowledge; (2) identification of gaps in that knowledge; (3) identification of work in progress; (4) evaluation of work as it relates to the identified gas; (5) development of conclusions; (6) formulation of recommendations for action by appropriate public and private agencies; and (7) estimation, when appropriate, of the benefits that might accrue through implementation of the recommendations. Within this framework, functional areas were addressed as they relate to specific situations; end uses were considered when fire was a design consideration and the end uses are of concern to the sponsors of the study.

Attention was given to natural and synthetic polymeric materials primarily in terms of their composition, structure, relation to processing, and geometry (i.e., film, foam, fiber, etc.), but special aspects relating to their incorporation into an end-use component or structure also were included. Test methods, specifications, definitions, and standards that deal with the foregoing were considered. Regulations, however, were dealt with only in relation to end uses.

The products of combustion, including smoke and toxic substances, were considered in terms of their effects on human safety; morbidity and mortality were treated only as a function of the materials found among products of combustion. The question of potential exposure to fire-retardant polymers, including skin contact, in situations not including pyrolysis and combustion were addressed as deemed appropriate by the committee in relation to various end uses.

In an effort to clarify the understanding of the phenomena accompanying fire, consideration was given to the mechanics of mass and energy transfer (fire dynamics). The opportunity to develop one or more scenarios to guide thinking was provided; however, as noted above, firefighting was not considered. To assist those who might use natural or synthetic polymers in components or structures, consideration also was given to design principles and criteria.

In organizing its work, the committee concluded that its analysis of the fire safety of polymeric materials should address the materials themselves, the fire dynamics situation, and the large societal systems affected. This decision led to the development of a reporting structure that provides for separate treatment of the technical functional aspects of the problem and the aspects of product end use.

Accordingly, as the committee completes segments of its work, it plans to present its findings in the following five disciplinary and five end use reports:

Volume 1	Materials: State of the Art
Volume 2	Test Methods, Specifications, and Standards
Volume 3	Special Problems of Smoke and Toxicity
Volume 4	Fire Dynamics and Scenarios
Volume 5	Elements of Polymer Fire Safety and Guide to the Designer
Volume 6	Aircraft: Civil and Military
Volume 7	Buildings
Volume 8	Land Transportation Vehicles
Volume 9	Ships
Volume 10	Mines and Bunkers

Some of the polymer applications and characteristics are in the classified literature, and the members of the committee with security clearance believed that this information could best be handled by special meetings and addendum reports to be prepared after the basic report volumes were completed. Thus, the bulk of the output of the committee would be freely available to the public. Considering the breadth of the fire safety problem, it is believed that exclusion of classified information at this time will not materially affect the committee's conclusions.

1.2 Scope and Limitations of This Report

This volume contains an overview of the contents of Volumes 1 through 4, which are the disciplinary documents of this series on the Fire Safety Aspects of Polymeric Materials. It is neither as complete nor as detailed as Volumes 1 through 4.

Volumes 6 through 10 are user documents and, as such, it was felt a condensation of the information contained therein would be of minimum use to the reader in search of data relevant to a particular field in the fire safety area. Most tables, graphs, drawings, photographs, and references have been omitted; however, all subject matter is covered in the same sequential order as in the complete disciplinary volumes so that information augmentation may be accomplished rapidly and with a minimum of inconvenience.

Conclusions and recommendations for each disciplinary document are included in their entirety in each related chapter of this volume; however, because of the condensed nature of the material, the conclusion rationale may not be as apparent or as completely defined as in each disciplinary volume.

If the reader has a requirement for fuller and more detailed information than is contained within this volume, it is recommended that the corresponding specific disciplinary volume be utilized for the additional data.

CHAPTER 2

MATERIALS

2.1 Introduction

The fire safety suitability of a material for a particular application depends on many factors, e.g., ease of access, ease of egress, proximity of the ignition hazard, proximity of other materials, duration of the fire, the thermal flux generated, the source of ignition, the ambient oxygen partial pressure, as well as the fire and smoke detection and suppression systems in place.

Development of polymeric materials with improved fire safety characteristics is a process in which competing requirements must be reconciled, because the improvement must be achieved at acceptable cost without impairment of those physical properties for which the polymer is selected, and with attention to the proper balance of the fire safety characteristics.

No organic polymeric material can withstand intense and prolonged heat without degradation, even in the absence of oxygen. Given sufficient oxygen and energy input, all commercial polymeric materials will burn. The terms: fire safe, fire proof, fire resistant, fire retarded, flame resistant, flame retardant, etc., express degrees of resistance of a material to burning, but most terms in current use lack precise meaning.

In this study the terms have the following meaning. Fire safety means primarily immunity of humans from damage caused by a fire; a fire retardant is a substance or treatment that reduces the flammability of a material; a fire retarded polymer is one whose flammability has been decreased by the incorporation of a fire retardant. Flammability refers to ease of ignition.

Essentially two avenues are open to the preparation of polymers with improved fire performance. The first consists in the development of new polymers whose fire safety characteristics are inherent in their structure. Although some materials in this class are available commercially now, this approach is still largely experimental.

The second approach consists of improving the fire safety characteristics of available lower cost materials by adding fire retardants. At present this approach is technologically and commercially the more important of the two avenues. It may take the form of the application of a coating to the surface of the material, or the incorporation of a fire retardant into its bulk at some appropriate stage of processing.

Use of coatings to protect flammable substances is one of the oldest methods of fire retardation. In this approach, a fire retardant coating is either applied to the surface of the material (nonintumescent coating) or produced in the presence of the flame (intumescent coating).

Polymers may be fire retarded by partial replacement of a conventional monomer in a polymer chain with a specific fire retardant monomer and/or by introducing into the bulk of the material certain fillers, e.g., hydrated alumina, or by introducing certain compounds with fire retardant properties. The first two methods lower the fire load, by heat absorption or by providing an inert diluent for the fuel. Fire retardants are usually halogen, phosphorus, nitrogen, antimony, or boron compounds, and can be used in combination. Synergism is a common effect.

According to the Report of the National Commission on Fire Prevention and Control (1973), fire claims nearly 12,000 lives annually in the United States. The report estimates the annual U.S. fire cost at 11.4 billion dollars.

Cost is an important aspect of fire retardation. Many thermally stable polymers with superior fire safety characteristics are currently expensive, in fact, too expensive for routine use in all but special applications in which the high performance requirements justify the cost. Advances in production technology and increases in consumption may be expected to lower costs, and, in some cases, to allow some of these polymers to be more widely used.

Application of fire retardant coatings can be a cost-efficient approach. However, it is limited to a relatively small number of uses.

The cost of fire retardation by the incorporation of a fire retardant in the polymer varies greatly, according to the performance level desired, the particular compound or treatment used, and the offset in fabrication costs. When it can be incorporated as a part of the production of the polymer it can be more economical than when it is used as an after-treatment.

Some halogenated organic compounds, organophosphorus compounds, metal salts and other inorganic substances, like antimony oxide and ferrocene, may accumulate in the environment, concentrate in food chains, and produce a variety of health and ecological effects. Skin contact, contamination of foods and inhalation can occur whether combustion is involved or not.

During its deliberations the committee was often presented with clear indications that many unnecessary fire hazards result from a general lack of knowledge and appreciation of fire safety problems in the use of polymeric materials. This lack of knowledge by the general public is understandable, if no less tragic in its consequences. Many uses of synthetic polymers are relatively new and no "conventional wisdom" has yet developed concerning their fire performance. Many synthetic polymer materials burn differently from the more familiar natural ones. They may melt and drip, and often give off dense and acrid smoke.

It is more difficult to justify the lack of understanding of polymer properties in general, and their fire safety aspects in particular, among practicing engineers, architects, designers, and builders. Quite generally, professionals who routinely recommend and specify the use of polymers are without even rudimentary training in polymeric materials. Undoubtedly, the general lack of understanding of the nature and properties of polymers results from the fact that polymer science and engineering are only 30–40 years old, and acceptance of the need for training in these relatively new materials in schools has been generally slow, both at the secondary level and tertiary level.

In the course of its regular activities, the committee listened to presentations by the sponsors and other qualified groups describing their problems and programs in the area of fire safety aspects of polymeric materials. On the basis of this input, it became apparent to the committee that planning, coordination, and dissemination of these efforts could be improved. The committee is aware of the establishment of the National Fire Prevention and Control Administration under the Act of 1974, and recognizes that the frame of reference under which this Administration will work, is currently being clarified. The committee is also aware of major ongoing programs at the National Bureau of Standards Center for Fire Research and in various committees of the American Society for Testing Materials.

The committee recognized that extensive research aimed at the development of products with improved fire performance is being conducted in industrial organizations. Information on much of this ongoing research is considered proprietary, and is not discussed in this report.

The committee senses a general need for better communication and wider dissemination of results in the fire research community, comprising government agencies as well as university and industrial research laboratories. A number of conferences are now being held on fire problems. Specialized journals are appearing and abstracting services and data banks are being developed, but there is some concern with the fact that there is still no general access to potentially important technical information. The committee recognizes that the National Fire Prevention and Control Administration is planning the establishment of a National Fire Data Center (Tabor, 1975). It is hoped that its existence will improve the dissemination of research results.

2.2 Wood and Wood Products

The appearance, texture, weight, cost, insulation characteristics, ease of fabrication and availability of wood and wood-base products have long contributed to their wide use in housing and building construction in the United States. While the combustibility of untreated wood has limited its application in some types of building, the use of proper fire safety design (for such items as heating systems, electrical systems, and exits), the combination of wood with fire retardant materials, and the slow charring rate and good strength retention of large wooden members in fire situations, have led to the general acceptance of untreated wood as a relatively safe building material.

Considerable technical information is available on methods for treating wood to reduce its flammability as well as initial rate of heat release and to render it self-extinguishing of flame and glow.

2.2.1 Composition and Pertinent Properties of Wood

The cellular structure of wood is complex. Cells are bound together by cementing layers. The cell walls are composed of cellulose, from 18 to 25 percent (dry wood weight) lignin, and, depending upon species and growth conditions, from 4 to 20 percent (dry wood weight) of other materials including aliphatic and aromatic hydrocarbons, alcohols, acids, aldehydes, esters, ethers, tannins, fixed oils, resin acids, proteins and alkaloids.

Some pertinent properties of wood are listed in Table 1.

Table 1. Some Properties of Wood

Property	Measure
Specific Gravity (most commercial species in the United States)	0.3 to 0.7 g/cm^3
Thermal Conductivity	
Perpendicular-to-grain (S = specific gravity; M = percent moisture)	[S(4.78 + 0.097M) + 0.567] X 10^{-4} cal/sec/cm^2 /°C/cm
Parallel-to-grain	Approximately 2.5 times the thermal conductivity perpendicular to grain
Specific Heat	0.34 cal/cm^3
Thermal Diffusivity	13 to 19 X 10^{-4} cm^2 /sec
Total Heat Value	4,600 to 4,800 cal/g
Charring Rate (ASTM E119 exposure, average species)	0.065 cm/min, decreasing with increasing specific gravity and moisture content
Rapid Surface Charring	280°C
Rapid Ignition Temperatures*	
Pilot Flame	225°C (10 min) to 300°C (1 to 2 min.)
Spontaneous	360°C to 540°C
Flame Spread Classification (ASTM E84)	60 to 80 for redwood, hemlock, northern spruce, Douglas fir; up to 230 for certain pines
Smoke Developed Classification (ASTM E84)	100 to 300
Smoke Density (various species) (Thickness: 0.50 in.; maximum specific optical density)	50 to 150 flaming; 250 to 550 non-flaming
Limiting Oxygen Index	20.5 to 22.7

*The ignition temperature of wood depends on specimen size, heating conditions, time, and test methods. Welker (1970) discusses some of the parameters involved in the measurement of the ignition temperature of wood.

2.2.2 Fire Retardant Treatment of Wood

In the early 1890s, the first U.S. patents on fire retardant treatments for wood were granted and the first commercial treatment began in New York City. Since then, the use of these treatments has fluctuated considerably. A peak was reached in 1943 when more than 5 million cubic feet of lumber and plywood were treated.

Much of this treated wood was used in building hangars for dirigibles. In 1957, only 362,000 cubic feet were treated annually with fire retardants. Increasing recognition by authorities of the benefits of treatment caused more than a tenfold increase over the next ten years; in 1966 almost 5 million cubic feet of lumber and plywood were treated. This level of treated wood production remained approximately the same for the next four years, but increased to almost 6 million cubic feet in 1975. To put this latter figure in perspective, it must be noted that this amount represents less than 0.1 percent of the wood and plywood products produced annually in the United States.

As noted above, the cost of treating wood with fire retardants has been a principal reason for its limited use. Although the chemicals generally are inexpensive, the cost of the treatment process increases the selling price of wood and wood products by 50 to 100 percent.

Fire retardants for wood are of two general types: (1) those that are impregnated into the wood or incorporated into wood products in the form of aqueous solutions, and (2) those that are applied as paints or surface coatings. Impregnation is most common, particularly for new materials, and the use of coatings has been limited primarily to materials in existing construction.

Fire retardants are incorporated into wood under pressure. The treatment involves exposing wood to a vacuum in an autoclave, introducing a 12 to 18 percent aqueous solution of the retardants and subsequently applying air pressure. Following impregnation, the treated wood must be dried under carefully controlled conditions to prevent degradation, darkening, staining and grain raising. Simple surface treatment usually does not result in the required retention.

Most of the chemicals used in fire retardant formulations have long been known. These chemicals include ammonium phosphate, ammonium sulfate, zinc chloride, borax and boric acid. A typical fire retardant formula consists of 35 parts of zinc chloride, 35 parts of ammonium sulfate, 25 parts of boric acid and 5 parts of sodium dichromate.

In the formulation of fire retardant treatments, consideration must be given to their effect on the strength, durability, hygroscopicity and corrosion properties of wood, as well as to fabrication characteristics (ease of painting, gluing and machinability) of the treated wood. Reduced performance in some applications has limited the use of fire retardant treated wood-based products; however, such problems can be minimized through use of appropriate chemical formulations and other techniques.

Proper treatment of wood and wood-based products with fire retardant formulations decreases the rate of surface flame spread and renders them self-extinguishing of flame and glow when the external source of fire or heat is removed.

There are a number of theories about how fire retardant chemicals act to reduce the flammability of wood; they may be summarized as follows:

1. Fire retardant chemicals form a liquid or glassy layer that prevents escape of flammable products and restricts the access of air to the combustion zone.

2. The chemicals form coatings, glazes, or foams that insulate the wood surface and prevent pyrolysis.

3. The treatments increase thermal conductivity of the wood thus permitting heat to dissipate from the surface at a rate faster than it is supplied by the ignition source.

4. Fire retardant chemicals undergo chemical and physical changes that absorb enough heat to prevent wood surfaces from reaching ignition conditions.

5. Nonflammable gases released by decomposition of the fire retardant chemicals dilute the combustion gases formed by pyrolysis of the wood and form a nonflammable gaseous mixture.

6. Fire retardant chemicals release certain free radicals capable of breaking the normal combustion chain, thus restricting the flammability range for combustible gas-air mixtures.

7. Fire retardant chemicals lower the temperature at which pyrolysis starts, thus promoting greater char yields amd reducing the formation of intermediate combustion products such as flammable tars and gases.

Recent research using thermogravimetric analysis, differential thermal analysis and product analysis has tended to support the last theory.

2.2.3 Wood Products

In addition to lumber and plywood, many other wood products are fabricated from wood chips or fibers. Of these products, panel board and paper are the most important.

Fire retardant treatments for paper and wood-based panel board may employ the same chemicals used in treating wood and plywood. The chemicals are often added during the fabrication process. Because many effective fire retardants are water soluble, it is difficult to retain enough chemicals in the mat produced through wet forming (although some fabricators have developed "closed wet forming" systems to partially solve this problem). When it is not possible to add the chemicals during the wet-forming process, they must be applied to the mats after forming. In the dry-forming process for making boards, the chemicals in either spray solution or fine powder form can be added to the fibers or chips before forming, but care must be taken to prevent the chemicals from migrating to the lower side of the mat.

Fire retardant chemicals added during fabrication must be properly selected to ensure that they do not interfere with resin bonding or board formation. Lack of consideration of this point may result in reduced strength and dimensional stability or contribute to degradation during final hot press forming operations.

2.3 Fibers

2.3.1 Introduction

Because fibers form the main components of clothing and furnishings, their flammability characteristics are of major concern.

Fibers are discussed under the heading of natural fibers, synthetic fibers, glass fibers, and fibers from relatively thermally stable polymers.

In textiles, as in many other applications of polymers, fire retardance is generally evaluated by laboratory tests that range from pragmatic simulation of use conditions to research methods designed to measure small differences and trends in the relative flammability of the materials. An often used test method is the Limiting Oxygen Index (LOI). This measures the percentage of oxygen in an oxygen-nitrogen mixture which will just support burning under specified conditions. A low LOI indicates high flammability. A comparison of the LOI values of some important textile materials is presented in Table 2; it shows that the LOI values of non-fire-retardant fibers may vary by about a factor of 1.5. Since the oxygen content of air is about 21 percent, a fiber with a substantially higher LOI value will not burn in air when ignition is attempted with the material in a configuration similar to that of the test. LOI values categorize materials only within the limits of those conditions under which the test is conducted (thermal flux, mode of ignition, air flow, sample size, sample shape, smoke orientation, substrate, etc.). Thus, the test results cannot be used as an absolute measure of the fire performance of any material.

Table 2. Limiting Oxygen Indices of Spun Fabrics (Fabric Weights: 4.8 to 7.0 oz/yd²)

Fabric	LOI (% O_2)
Acrilan (acrylic)	18.2
Arnel Triacetate (cellulose ester)	18.4
Acetate (cellulose ester)	18.6
Polypropylene (polyolefin)	18.6
Rayon	19.7
Cotton (greige)	20.1
Nylon	20.1
Polyester	20.6
Wool (dry cleaned)	25.2
Dynel (modacrylic)	26.7
Rhovyl "55" (polyvinyl chloride)	27.1
Nomex N-4272 (aromatic polyamide)	28.1

2.3.2 Natural Fibers

Cotton is the most extensively used natural fiber; chemically it is essentially cellulose and, therefore, rich in fairly reactive hydroxyl groups. In principle, cotton, in the form of fiber, yarn, or fabric, can be treated with fire retardants; in practice, fire retardant treatment of the finished fabric is the only commercially significant process.

Durable fire retardants are defined as those that can provide the desired degree of fire retardance for the useful life of a textile product. This means durability over 50 or more laundry cycles. Other important criteria include strength retention, avoidance of unwanted stiffness and absence of discoloration of the treated material. Criteria that can sometimes disqualify a finish are odor, sensitivity to acids or bases, and ion exchange properties. Currently, acceptable durable fire retardants for cotton and rayon fabrics are of two types: metal oxides and organophosphorus compounds.

Two techniques have been proposed for using metal oxides in laundry-durable finishes:

1. Formation of an insoluble titanium-antimony oxide compound within the fiber, with possible chemical attachment to the cellulose.

2. Binding antimony oxide and a chlorocarbon to the fiber with a nitrogenous resin.

Organophosphorus compounds penetrate the fiber. They react, polymerize, or copolymerize with an appropriate monomer or, in some systems, with the cellulose.

Another way of applying organophosphorus compounds is by depositing preformed phosphorus-containing polymers on the fibers or fabrics. Subsequently, these are either further polymerized or fused to provide durability. Substitution reactions without polymer formation, such as phosphonomethylation, also have been used to impart fire retardance, but with limited commercial success.

Approaches that have been considered for commercialization, or have actually been implemented on a commercial scale, are the following:

Tetrakis (hydroxymethyl) phosphonium chloride (THPC)

$$(HOCH_2)_4\ P^{+}Cl-$$

is a white water-soluble crystalline compound: The corresponding quaternary acetate, sulfate, oxalate and phosphate salts can be prepared by substituting the corresponding acid for hydrochloric acid in the reaction with phosphine and formaldehyde.

A fire retardant for medium and heavy weight cotton fabrics, based on THPC, urea, methylolmelamine and various textile modifiers, has been in use since about 1959 (primarily in Europe). It is produced by impregnating cellulosic fabrics with the product obtained by reacting urea with THPC and then insolubilizing this compound with ammonia. The most recent THPC-based fire retardant process to become commercialized in the United States is produced by impregnating cellulosic fabrics with partially neutralized THPC and then forming a fire retardant polymer inside the fibers by reacting the methylol phosphorus compound with ammonia. This is an unusual retardant in that the chemical treatment does not reduce the strength of the treated fabrics.

Another organophosphorus compound, not used currently because of physiological hazards, is tris(1-aziridinyl)-phosphine oxide (APO):

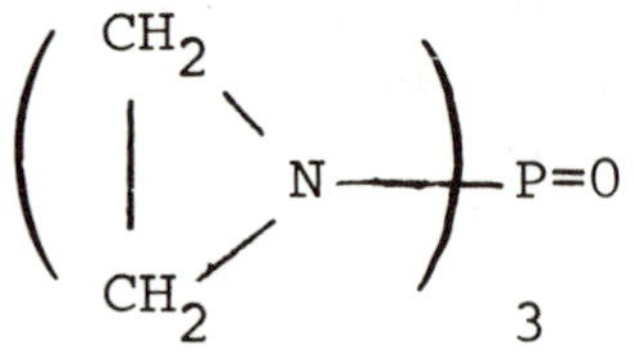

$$\left(\begin{array}{c} CH_2 \\ | \quad \rangle N - \\ CH_2 \end{array} \right)_3 P{=}O$$

Used alone it can impart fire retardance and a high degree of wrinkle resistance to cellulosic fabrics.

Reactive phosphonates, insolubilized on fabric either by reaction with cellulose or *in situ* polymerization, have also been proposed as durable fire retardants for cotton.

In general, good fire retardance is obtained on cotton fabrics through insolubilization of about 2 to 3 percent phosphorus, preferably in conjunction with nitrogen.

Wool is basically a sulfur containing protein (keratin). Wool textiles are generally less flammable than cellulosics, and less widely used. Thus, only limited studies are available on methods to enhance fire retardance in wool textiles. The flammability of wool generally is decreased by treatment with organophosphorus compounds or with specific salts of polyvalent metals.

2.3.3 Synthetic Fibers

The production of man-made and natural fibers in the United States in 1973 is summarized in Table 3. All of these fibers either ignite and propagate flame, or melt and shrink (thus exposing the wearer directly to flame), or form a friable char that afford no effective protection.

System design of a textile construction is often effective in reducing fire risk. For instance:

1. Carpets of unmodified acrylic, nylon or polyester can be made to resist ignition by a cigarette, fireplace ember, or the standard methenamine fuel tablet, by: a) increasing the weight of the face fiber by a few ounces per square yard; b) using hydrated alumina as a heat absorbing filler in the carpet backing; or c) applying a heavier and more even coating of backing adhesive.

2. When unmodified thermoplastic fibers resist ignition, it is because they melt away from the ignition source. To take advantage of this phenomenon, one must: a) avoid using these fibers in proximity to non-melting fibers (e.g., cotton, rayon, or acrylic); b) avoid the use of non-melting sewing thread; c) determine whether the fabric, given a specific construction and weight, will drip out upon ignition.

Table 3. Production of Natural and Man-Made Fibers in the United States in 1973

Fiber	Amount Produced (billion lb)
Natural	
Cotton	3.65
Wool	0.17
Man-Made	
Rayon	0.89
Acetate	0.46
Nylon	2.18
Polyester	2.77
Olefin	0.42
Acrylic	0.74
Glass	0.69
Other	0.13

Chemical modifications are introduced to promote fire retardance by:

1. Altering the course of the thermal oxidative degradation of the polymer that supplies the volatile fuel to ignite and propagate flame.

2. Emitting free-radical-trapping and/or energy absorbing fragments into the volatile fuel, thus cooling or quenching the flame reactions.

3. Accelerating the decrease of viscosity of the polymer at the flaming surface, thus promoting drip-out.

Chlorine, bromine, phosphorus, nitrogen, sulfur, boron, and antimony play dominant roles in the chemical modification of flammability properties of polymers. These flammability-modifying elements can be incorporated into fibrous structures in many ways, including:

1. Monomeric building blocks containing one or more of these elements can be incorporated directly in the polymer chain (e.g., vinyl bromide can be copolymerized into a modacrylic).

2. The polymer can be modified by grafting (e.g., vinyl chloride can be grafted onto preformed poly(vinyl alcohol)).

3. An additive can be incorporated into the polymer solution or melt before extrusion into fiber form (e.g., hexapropoxy phosphotriazene can be introduced into viscose dope prior to extrusion to form rayon).

4. The textile material (fiber, yarn or fabric) can be treated topochemically with reagents from solutions or emulsions (e.g., brominated alkyl phosphate may be sorbed onto polyester, or THPC with a coreactant onto rayon).

Fire retardant variants exhibit different durabilities to repeated laundering. Important factors in laundering are water hardness, the chemical nature of detergent and soap formulations, bleaches and exposure to sunlight.

Both regenerated cellulose and cellulose acetate derived from cellulose, a natural polymer, are best included among the synthetic fibers. The term "regenerated cellulose" describes cellulose that has been dissolved in the form of a soluble derivative and subsequently reprecipitated.

Cellulosics

Unmodified rayon ignites readily and burns rapidly to completion Fire retardant treatments for rayon encompass all the means available for treatment of cotton. In addition, it is possible to incorporate effective additives into the viscose spinning solution prior to extrusion. One such additive, used in a commercial product as noted above, is an alkoxy phosphazene that remains dispersed in the fiber and shows excellent durability.

Appreciable quantities of nonwoven, disposable textile items are produced from rayon. Because they are not laundered for reuse, it is possible to achieve fire retardance with approximately 20 percent loading of water soluble salts (e.g., ammonium sulfamate or diammonium phosphate).

The unmodified cellulose ester fibers (mainly cellulose acetate and triacetate) ignite and propagate flame. The fabric drips while continuing to burn.

The principal method used to achieve fire retardance in cellulose esters is to incorporate a brominated alkyl phosphate ester into the spinning solution before extrusion. The fire-retarded product either resists ignition or the affected portion melts and drips out at once.

Polyesters

The most important polyester fiber is poly(ethylene terephthalate). While the fiber is flammable, many polyester textiles will not ignite because the fabric melts away from a small ignition source or self-extinguishes by drip out. Both of these mechanisms will fail with heavier constructions or if even only a small amount of non-thermoplastic fiber (e.g., cotton) is present.

To improve fire retardance in heavier weight constructions, it is possible to use several approaches. For example, the molecular composition can be altered to contain bromine by using 2,5-dibromoterephthalic acid to replace some terephthalic acid, or by using bis(hydroxyethoxy)tetrabromobisphenol A to replace some of the ethylene glycol. Antimony oxide then can be included in the formulation to enhance the effectiveness synergistically.

The comfort and aesthetics of poly(ethylene terephthalate) can be enhanced significantly by blending it with cellulose acetate or triacetate. When cellulose acetate contains appropriate fire retardants, or when fire retardants are topochemically applied to the blend fabric, good fire retardance can be obtained for some fabric constructions.

Nylon

Nylons are polyamides. The unmodified polyamides, nylon 6, nylon 66, nylon

610, etc., perform essentially like unmodified polyester with respect to flammability. Nylons resist ignition and flame propagation in vertical configurations because of drip-out, except in heavier weight fabrics. Carpets with nylon face fiber readily pass the methenamine fuel test with even less face fiber density than polyester. Nylon blends with cellulosics represent a flammability hazard. Published means for improving nylon fire retardance have serious deficiencies. Each method impairs some other desirable attribute such as performance, durability or economics.

Acrylics and Modacrylics

Acrylics are polymeric fibers containing at least 85 percent poly-acrylonitrile. Modacrylics are copolymers of acrylonitrile with vinyl chloride, vinylidene chloride, and/or, possibly, other vinyl monomers.

Acrylics shrink away from a small ignition source and therefore escape ignition. Once ignited, however, acrylics burn vigorously, ejecting plumes of flaming gases and dense smoke. Acrylics are made acceptable for the carpet pill test by blending the face fiber with 20 percent of modacrylic fiber.

Modacrylics suffer from certain shortcomings in textile performance. Historically, they have a low softening point and poor thermal dimensional stability although some improvements have been made. These fibers find acceptance in children's sleepwear because they are relatively inexpensive, resist ignition, and burn slowly or self-extinguish if ignited.

Poly(vinyl chloride) (PVC) fibers as such can be considered to be inherently and permanently fire retarded by virtue of their high chlorine content. As a textile fiber, PVC has exhibited poor aesthetics and poor dimensional stability, but an improved heat- and solvent-resistant syndiotactic form of PVC has been developed in Italy.

Saran filaments offer high strength and flexibility and are chemically resistant and self-extinguishing. The monofilaments have been widely used in automotive seat covers, outdoor furniture, agricultural shade cloth, filter fabrics, insect screening, window awning fabrics, venetian blind tape, and brush bristles. Extruded multifilament yarns have been used for draperies, upholstery, doll hair and other wigs, dust mops, and various industrial fabrics.

Polyolefins

Polyolefin fibers are polyethylene or polypropylene. These fibers ignite readily and continue to burn with flaming drops. No successful formulas are known that retard the flammability of polyolefin fibers.

2.3.4 Glass Fibers

For some end uses, glass fibers are important. They melt at about 815°C but do not burn. They are generally treated with organic finishes to enhance their resistance to abrasion and to improve other functional properties. The flammability hazard of glass fibers in actual use is significantly increased by the presence of these surface modifying organic materials. Fabrics made from glass fibers are used widely in various industrial applications, in composites and laminates and in draperies and curtains.

2.3.5 Fibers From Thermally Stable Synthetic Polymers

An important approach to fire retardant synthetic fibers will be the utilization of some relatively new classes of thermally stable synthetic polymers with backbones containing nitrogen.

At this time, many of these polymers are expensive. Regulations can be expected, however, to require the use of flame resistant textiles in many additional areas including mass transit. It is probable that modern chemical technology will be called on to provide lower cost approaches to thermally stable polymers.

Aromatic Polyamides (Aramids)

This composition is the basis for the first commercially available thermally stable aromatic polyamide fiber. Nomex (poly(*m*-phenylene isophthalamide) is duPont's first version of this class of polymers; it decomposes at about 370°C. In low-thermal environments involving a small ignition source, Nomex is self-extinguishing; however, when exposed to a large heat flux, it will shrink, burn, and propagate flame to other materials. An improved version, HT-4, ranks with the best of the thermally stable fibers. A comparison of LOI values is presented in Table 4.

Table 4. Limiting Oxygen Indices of High Performance Fabrics (Fabric Weight: 4.5 – 7.0 oz/yd²)

Fabric	LOI (%O_2)
Untreated Cotton	16-17
F. R. Treated Cotton	31-32
Natural Nomex	27-28
Dyed Nomex	25-27
Kynol	29-30
Fypro	29-30
Durette	35-38
PBI (polybenzimidazole)	38-43
Nomex HT-4	42-46
PBI-S	42-49

Nomex currently is being used in military clothing, race drivers' garments, hospital bedding, filtration bags, various forms of transmission belting, industrial protective clothing, and children's sleepwear. Staple Nomex formed into paper is being used as electric motor insulation and as honeycomb core for aircraft wall panels.

Potential future applications in aircraft interiors, parachutes and carpeting depend on improving color fastness.

The development of a more recent aramid, Kevlar, has been directed toward the production of high modulus fibers for use as reinforcement in structural composites and tires. Tenacities in the 19–22 g/d range, with a rupture elongation of 3 to 4 percent, have been reported for Kevlar and the same tenacity with 2 to 3 percent elongation for Kevlar 49. This tremendous gain in tenacity, coupled with densities of 1.40 to 1.45 g/cm^3, provides a potential replacement for glass, steel, nylon, and polyester fibers now used as tire cord, provided the new materials can be made at a competitive cost.

Polybenzimidazole Fiber

Polybenzimidazole (PBI) is prepared from a diester of isophthalic acid and diaminobenzidine and is spun from a solvent using typical dry spinning techniques. The initial objective was to obtain a material having long-term thermal stability at temperatures above 500°C, but extensive evaluation of fiber, yarn and fabric demonstrated that PBI did not have the desired long-term thermal stability. Measurements of burning rate in air and oxygen indicate that this fiber will not burn at 250–300°C. This characteristic, in combination with high moisture regain (13 percent compared to 8.5 to 10.3 percent for cotton) has resulted in extensive studies of PBI for clothing and related applications where comfort and fire safety are important requirements.

Aromatic Polyimides

Aromatic polyimides are synthesized from an aromatic dianhydride and an aromatic primary diamine. They are used primarily for moldings, composite matrix resins, films and foams. The films are the best non-halogenated fire resistant films available. They are clear, amber colored, fairly flexible, and have high tensile strengths. Fibers have been spun from polyimides and their mechanical properties have been investigated exhaustively. A typical polymide repeat unit is poly(oxydiphenyl pyromellitimide) (Kapton®).

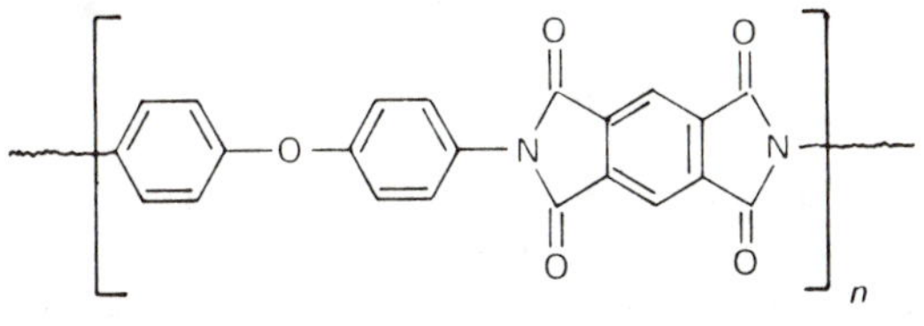

A variety of thermally stable, high-char-forming heteroaromatic polymers have been studied, primarily as matrix resins for composites. A number of these polymers have been formed into fibers. Polyoxadiazole (PODZ) has been formed by dry spinning the polyhydrazide precursor polymer from dimethyl sulfoxide followed by thermal cyclodehydration to form the final structure.

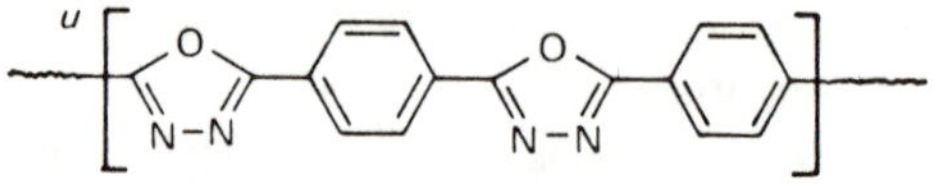

Although long-term thermal stability of these fibers was superior to PBI, they were found to be extremely sensitive to light.

The most recent and most promising candidate in the high-temperature polymeric textile fiber field is poly (bisbenzimidazobenzophenanthrolinedione), labeled "BBB." When tested at 642°C the fibers retained 60 percent of their room temperature tenacity. Usable strengths may exist well above 697°C. This fiber is not flammable. Disadvantages at this time are primarily cost and color.

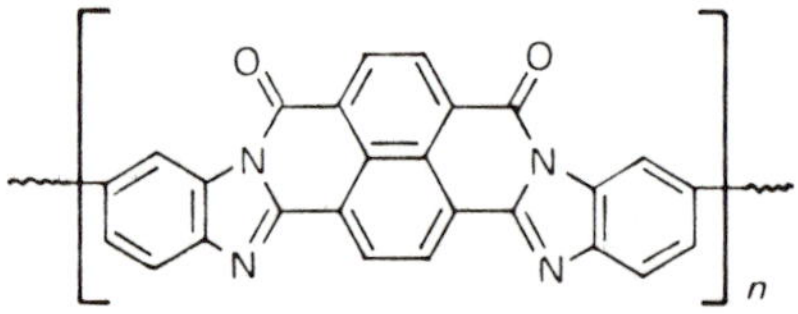

Only one phenolic fiber, Kynol®, has been developed. In its preparation a phenol-formaldehyde polymer is first spun into multifilament yarns and is then cured with formaldehyde. The Kynol® fiber is light yellow, but can be bleached to a pure white. The strength of the fiber is low, as is its abrasion resistance. When exposed to flame, the fiber chars on the surface, retaining its shape with no distortion. Its LOI is superior to that of Nomex® and the fiber is resistant to thermal shrinkage. Because of its poor abrasion resistance, Kynol® is blended with other fibers to effectively utilize its flame resistance and thermal stability. Felt and batting made of Kynol® have potential for providing both insulation from cold and protection from fire in cold weather garments. Although raw material costs are low, processing will be costly until the manufacturing process is scaled up to larger volumes.

2.3.6 Fire Safety Aspects of Fiber Blends

Fabrics made of yarns containing two or more fibers of different chemical compositions and properties have attained great commercial importance in textile markets. The technology of these blends has become an important illustration of the textile industry's skill in the optimal utilization of available fibers for the manufacture of new or improved products. The fire safety aspects of blend fabrics have received attention only since about 1969, as a consequence of regulations and industry wide activity on general textile fire safety problems.

Fiber blends pose special fire hazards. For example, care must be taken to avoid the presence of nonthermoplastic components in any blend of thermoplastic fibers. Even cotton sewing thread should be avoided because it may serve as a wick to stabilize the flaming melt. Early investigations have established that the fire safety aspects of blends cannot be predicted from a knowledge of the behavior of individual fiber components. In effect, a blend becomes a new chemical entity with flammability properties of its own.

Fire retardant systems that are effective on a specific fiber do not necessarily provide useful approaches to the treatment of blends. Physical and chemical interactions of different fibers in blends under conditions of burning pose complex problems that are not understood and have not been studied adequately. However, a considerable amount of empirical work has been carried out during the last few years on the flammability of blends.

The flammability of fabrics made from blends of the aromatic polyamide Nomex® with some combustible fibers has been examined experimentally. Fabrics made from blends of Nomex® with fire retardant fibers generally exhibit high LOI values and self-extinguishing behavior in vertical fabric flammability tests.

Some data are available for blends of Kynol® with other fibers. For example, the relationship of oxygen index to Kynol® content for Kynol® wool blends indicates that "nonflammable" mixtures are obtained for Kynol® contents about 35 percent.

Fire retarded fabrics for specific uses have been obtained from blends of wool with PVC fibers and with modacrylic fibers. Fabrics made from 60/40 wool/PVC yarn have been reported to exhibit satisfactory properties and to pass the vertical flammability test prescribed in the children's sleepwear standard.

2.3.7 Fire Retardant Compounds and Additives

Generally, the flammability of polymers can be decreased by altering the products of thermal decomposition so that:

1) Noncombustible gases are generated that dilute the oxygen supply and tend to exclude oxygen from the polymer surface.

2) Radicals or molecules from degradation of retardant chemicals react endothermally or interfere with chain reactions in the flame or substrate species.

3) The retardant decomposes endothermally.

4) Nonvolatile char or liquid barriers are formed that minimize oxygen diffusion to the condensed phase and reduce heat transfer from flame to polymer.

5) Finely divided particles reduce flame propagation by altering the course of gas phase reactions leading to less reactive radicals.

Fire Retardants Based on Phosphorus

The most important among these compounds have been used primarily for cellulosic fibers and are identified in Table 5 with their textile applications. These phosphorus compounds are insolubilzed through reaction with hydroxyl groups, polymerization, or copolymerization *in situ.*

Table 5. Reactive Organophosphorus Fire Retardants

Compound	Type	Major Applications
$(CH_3O)_2P(=O)CH_2CH_2CONHCH_2OH$	phosphonate	Cotton fabrics
$(RO)_2P(=O)CH=CH_2$	phosphonate	Cotton fabrics
$(HOCH_2)_4P^+ Cl^-$	phosphonium salt (THPC)	Cotton fabrics Wool fabrics
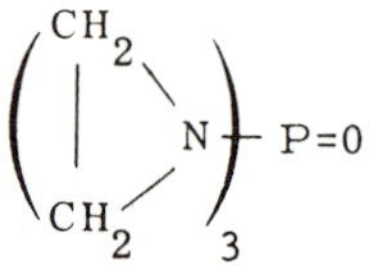 $(\text{aziridinyl}: CH_2CH_2N)_3P=O$	phosphoric triamide (APO)	Cotton fabrics
$(CH_3NH)_3P=O$	phosphoric triamide	Cotton fabrics
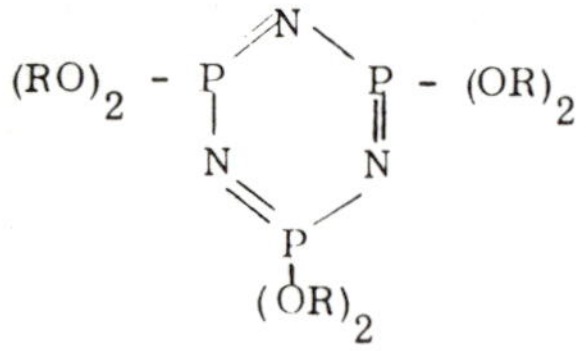 $[(RO)_2PN]_3$ (cyclic trimer: ring of alternating P and N, each P bearing $(OR)_2$)	phosphazene	Rayon fiber

Fire Retardants Based on Halogens

Halogen-containing fire retardants are used extensively for fibers and textiles in several ways. Halogenated monomers and comonomers are used in the manufacture of poly(vinyl chloride) fibers, modacrylic fibers, and poly(vinyl chloride) /poly(vinyl alcohol) bicomponent fibers. Typical of fire retardants in this class are: vinyl chloride, vinylidene chloride, vinyl bromide, and other halogenated vinyl monomers.

Halogenated compounds are believed to achieve their effectiveness primarily through vapor phase mechanisms that inhibit oxidative reaction in the flame. They are often used in conjunction with antimony trioxide, which interacts synergistically with the halogen. Important halogen-containing flame retardants used in fibers and textiles are summarized in Table 6.

Table 6. Halogen Containing Fire Retardants

Halogen Compounds	Mode of Use	Proposed Applications
Vinyl chloride $CH_2 = CHCl$	Monomer or comonomer in manufacture of fibers	PVC fibers Modacrylic fibers PVC/PVA fibers
Vinyl bromide	Comonomer in manufacture of fibers	Modacrylic fibers
2,5-Dibromo-	Comonomer in	FR polyester fibers
terephthalic acid $(Br)_2C_6H_2(COOH)2$	modified poly(ethylene terephthalate)	
Bis(hydroxyethoxy)-tetrabromobis-phenol A	Comonomer in modified poly(ethylene terephthalate)	FR polyester fibers
Tris(2,3-dibromo-propyl) phosphate	a) Additive in fiber manu-facture. b) Finishing of fabrics from 100% polyester.	FR acetate fibers FR Triacetate fibers FR polyester fibers
Chlorinated paraffin wax,	Fabric finishing	Industrial and military fabrics
PVC latex, PVB latex		

2.4 Elastomers

2.4.1 Introduction

Conventional elastomers consist of flexible linear chainlike molecules crosslinked (cured) to form a three-dimensional network. Newer thermoplastic elastomers are heterogeneous systems in which glassy or crystalline domains, interspersed in the rubber matrix, act as multifunctional crosslink points.

Elastomers are distinguished from other materials in being able to sustain very large (up to 700–1,000 percent) reversible deformations. They are essential in all applications when this property is required.

Most rubbers are compounded, i.e, they contain additives. These additives may be part of the curing system or may be added as reinforcement or for other special effects. In practical applications, elastomers are rarely used by themselves and are usually employed in combination with other materials, often with fibers and metals.

Elastomers that account for the bulk of rubbers used today are shown in Table 7, together with their approximate consumption in the United States in 1972.

Table 7. Consumption of New Rubbers in the United States—1972

Type	Amount Sold MM lbs
Styrene-Butadiene	3,189
Natural	1,411
Polybutadiene	682
Polychloroprene	273
Butyl (1)	268
Polyisoprene	250
Ethylene-Propylene-Diene	141
Nitrile (2)	141
Chlorosulfonated Polyethylene	50
Polysulfide	50
Silicone	50
Fluoroelastomers	50
Polyacrylates	50
Polybutene	50
Polysobutylene	50
Epichlorohydrin	<50

(1) Includes Chlorobutyl
(2) Includes Nitrile/PVC

Most organic elastomers burn easily when not fire retarded. Today, no elastomer has the desired combination of low flammability, low smoke emission, good mechanical properties, and reasonable cost. Prime areas of concern are coated fabrics, wire and cable coatings, hoses, and foamed elastomers.

2.4.2 Fire Safety Aspects of Elastomers

As with most organic polymeric materials, the incorporation of halogens, either as an additive or as an integral part of the molecule, decreases the flammability of elastomers. This is illustrated by polychloroprene, chlorinated polyolefins, epichlorohydrin rubbers, the various fluoro and chlorofluoro elastomers, halogen-containing polyurethanes, and various compositions in which halogenated additives are used. All these materials have deficiencies in their fire safety characteristics; they give off smoke and hydrogen halides on combustion and/or exposure to an intense fire environment.

As with other organic polymers, phosphorus compounds are used to decrease flammability.

A second method of improving the fire resistance of many polymeric materials is to promote char formation by structural modifications or by the use of additives.

The third approach to reducing flammability (and smoke) consists of replacing all or part of the carbon in the polymer structure with inorganic elements.

A further approach is to incorporate large amounts of inorganic fillers, which reduce the fuel value of a composition even if the fillers have no specific fire retardant properties. Fortunately, most elastomers can tolerate or even require a substantial amount (about 50 percent) of particulate inorganic filler. Alumina trihydrate is particularly effective because of the endothermic release of water.

2.4.3 Specific Elastomers

Hydrocarbon-Based Elastomers

This group includes natural rubber, synthetic cis-polyisoprene, poly-butadiene, styrene-butadiene rubber, butyl rubber, and ethylene-propylene rubbers as its main constituents.

Because these rubbers are low-cost materials with good mechanical properties, they are used in large volume; but they burn readily with much smoke. Fire retardant additives reduce flame spread and ease of ignition from low-energy ignition sources, but do not prevent burning in an intense fire situation. Alumina trihydrate is receiving intensive study as a filler to reduce flammability and smoke formation in these elastomers.

Chlorine-Containing Elastomers

These elastomers include polychloroprene, chlorinated ethylene polymers and copolymers (chlorinated polyolefins) as well as epichlorohydrin rubbers.

Polychloroprene has good mechanical properties, good oil resistance, is reasonably priced and widely used where a fire retarded elastomer is required. A wide variety of formulations of polychloroprene are available for various uses. The material has significantly better fire retardance than the straight hydrocarbon rubber. However, it generates large amounts of black smoke and hydrogen chloride gas when exposed to a fully developed fire.

Nitrile Rubbers

Nitrile rubbers are copolymers of butadiene and acrylonitrile. The ratio of butadiene to acrylonitrile is similar to the ratio of butadiene to styrene in SBR. The cyanide group imparts to these elastomers some properties of the halogen-containing rubbers, except for flammability. Because of their excellent resistance to hydrocarbons (superior to that of polychloroprene), they are extensively used in gasoline hoses and fuel tanks. Under fire conditions, the cyanide group constitutes a potential toxicological hazard (source of hydrogen cyanide).

Polyacrylate Elastomers

A series of polyacrylate elastomers are made by polymerization of esters of acrylic acid or by copolymerization of such monomers with other monomers. A typical polyacrylate elastomer is made from ethyl acrylate, and 2-chloroethyl vinyl ether.

Polyacrylate rubbers show excellent oil resistance at elevated temperatures (140°C). They are used for O-rings, gaskets, hose, oil seals, and wire insulation. These rubbers may be expected to be highly flammable.

Polyurethane Elastomers

Polyurethanes are polymers containing the group –NC–CO–0–. They are formed typically through the reaction of a diisocyanate and a glycol.

Polyurethane elastomers have outstanding mechanical properties, are moderately priced, and have unique fabrication capabilities. For these reasons they are used in a variety of rubber goods. Fire-retardant grades, generally based on bromine and/or phosphorus containing additives, are available, but they break down and burn in intense fires. Some generation is generally less than with hydrocarbon elastomers, but some hydrogen cyanide gas can be generated. The major use of polyurethane elastomers is in flexible foams.

Polysulfide Rubbers

These rubbers, also known as "thiokols," are polymers composed of aliphatic hydrocarbon chains connected by di-, tri-, and tetrasulfide links. Because of their outstanding resistance to hydrocarbons, they are used extensively as sealants in aircraft fuel tanks and pressurized cabins. Polysulfide rubbers are usually crosslinked by heating with zinc oxide or zinc peroxide, but other curing systems also are used. The flammability of a cured polymer may be improved by the incorporation of halogen, phosphorus, or antimony-containing compounds.

Fluorocarbon Elastomers

In these elastomers, hydrogen is replaced partially or wholly by fluorine. Viton® is essentially a copolymer of hexafluoropropylene and vinylidene fluoride. They may contain small amounts of another monomer to allow crosslinking. Fluoroelastomers, because of their high fluorine content, do not support combustion. They extinguish immediately upon withdrawal of a flame.

Specialty rubbers such as nitroso and triazine elastomers have been reported to be nonburning by the most rigid standards. Nitroso rubbers are copolymers of trifluoronitrosomethane with tetrafluoroethylene or (substituted) trifluoroethylene, represented by the following formula:

$$\left[\underset{\displaystyle CF_3}{\overset{|}{N}} - O \left(CF_2 - \overset{\displaystyle X}{\overset{|}{C}F} \right)_n \right]_m$$

in which X = H, F, or $CF_2 - COOH$.

Triazine elastomers are composed of triazine rings separated by perfluorinated polymethylene groups and carrying only perfluoroalkyl substituents:

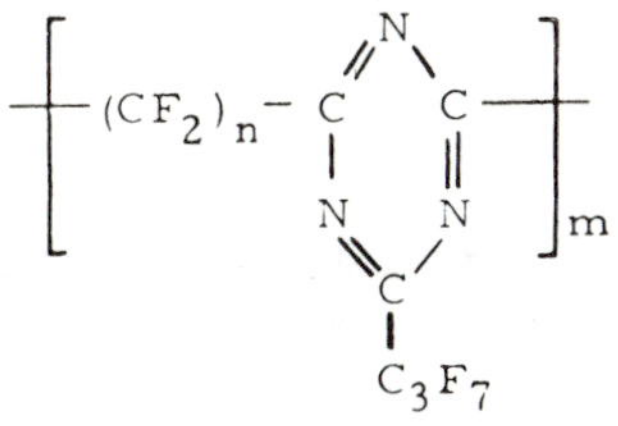

$$\left[-(CF_2)_n - \underset{\substack{|\\ N \\ \;\;\diagdown\!\!\diagdown \\ C \\ | \\ C_3F_7}}{C} \; \overset{N}{} \; C - \right]_m$$

These polymers have outstanding thermal stability and retain their elastomeric character after exposure to 260 to 430°C, but are still largely experimental.

Silicone Rubbers

Silocone rubbers typically contain the repeat unit where R is a hydrocarbon radical. The chain backbone of these rubbers is therefore inorganic. Finely divided silica is used as a reinforcing filler to obtain useful elastomeric properties. Cross-linking is usually achieved by organic peroxides or by polyfunctional silanes.

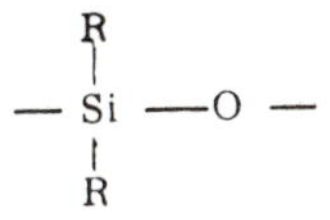

$$- \underset{\substack{| \\ R}}{\overset{\substack{R \\ |}}{Si}} - O -$$

Silicone elastomers generate relatively little smoke, they are fire resistant in air, and when burned, have low fuel values. They burn slowly and produce no flaming drip, but are relatively expensive. For many applications their mechanical properties are marginal. Silicones, however, offer promising combinations of fire safety aspects, physical properties and cost.

Phosphonitrilic Elastomers

Phosphonitrilic elastomers represent another example of "inorganic elastomers." The phosphorus-nitrogen backbone, supplies the flexibility required for elastomeric properties and contributes little fuel value.

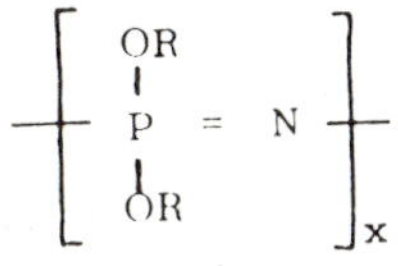

$$\left[- \underset{\substack{| \\ OR}}{\overset{\substack{OR \\ |}}{P}} = N - \right]_x$$

The phosphonitrilic materials are in early stages of development. Much needs to be done to define their utility and feasibility for various applications. For example, the combustion and pyrolysis products, contributed by the phosphorus and nitrogen, need to be characterized. Phosphonitrilic elastomers are, however, one of the principal present hopes for a low smoke, low flammability elastomer.

2.5 Plastics

2.5.1 Introduction

In this paragraph, the general fire safety aspects of various types of currently available plastics and their current applications are summarized.

Table 8 represents data on the U.S. sales volumes of the major plastics in 1973 and 1974.

Table 8. Plastics Sales in the United States in 1973 and 1974 (1000 metric tons)

Material	1973	1974
Acrylic	233	243
Alkyd	334	388
Cellulosics	77	76
Coumarone-indene and petroleum resins	160	160
Epoxy	102	106
Nylon	87	88
Phenolic	624	587
Polyester	468	425
Polyethylene (high density)	1,248	1,275
Polyethylene (low density)	2,691	2,769
Polypropylene	1,012	1,061
Polystyrene and styrene copolymers	2,356	2,328
Polyurethane	593	622
Poly(vinyl chloride) and other vinyl copolymers	2,151	2,180
Other vinyls	390	420
Urea and melamine	488	475
Others	138	147
Total	13,152	13,350

Table 9 presents Limiting Oxygen Index (LOI) values for several important plastics.

2.5.2 Approaches to the Fire Retardation of Plastics

Fire retardation may be achieved either by using an external coating or by incorporating fire retardants in the bulk of the material.

The use of fire retardant coatings to protect flammable substrates is one of the oldest of fire retardant methods. The coatings are of two forms, intumescent and non-intumescent, with the former being more useful. Intumescent coatings combine film-forming characteristics with char- or ash-forming and gas-producing capabilities. The thermally stable insulating char or ash protects the substrate from the thermal effects of the flame.

Table 9. Limiting Oxygen Indices of Various Plastics

Plastics	LOI (%O_2)
Polyacetal	15.0
Poly(methyl methacrylate)	17.3
Polyethylene	17.4
Polypropylene	17.5
Polystyrene	17.8
Poly(4-methylpentene)	18.0
Filter paper (cellulose)	18.2
ABS resin	18.8
Cellulose Acetate	19.0
Styrene-acrylonitrile	19.1
Poly(ethylene terephthalate)	20.0
Birch wood	20.5
Poly(vinyl fluoride)	22.6
Chlorinated Polyether (Penton)	23.2
Noryl 731	24.3
Nylon 66	24.3
Polycarbonate	24.9
Nylon 6	26.4
Poly(phenylene oxide)	30.0
Polysulfone, P-1700	30.0
Polyamides	31 - >45
Polyimide (Kapton film)	36.5
Polysulfone (PES)	38.0
Poly(phenylene sulfide)	>40.0
Poly(vinyl chloride)	40.3
Poly(vinylidene fluoride)	43.7
Chrolinated PVC	45.0
Polysulfone (Astrel)	>50.0
Poly(vinylidene chloride)	60.0
Polytetrafluoroethylene	95.0

Halogen Fire Retardants

Chlorine and bromine compounds are the materials most generally employed in polymer fire retardant applications; these materials have been used successfully to retard the burning of a wide variety of synthetic and naturally occurring polymers including wood, cellulose and cellulose derivatives, wool, polyethylene, polypropylene, other synthetic polyolefins, polyesters, polyurethanes, polyamides, and many other polymers. Their wide applicability has so increased their utilization that halogen compounds, usually used together with antimony or phosphorus now compose a market of more than 100 million pounds per year. They may be divided into additives and chemically reactive monomers. The additives are usually combined with the polymer during processing and do not react chemically with it. Reactive monomers are chemically combined with the polymer structure at some processing state.

Although halogen fire retardants are capable of imparting a high degree of fire retardance to polymers by themselves, their efficiency can often be increased considerably by the simultaneous use of selected compounds of boron and phosphorus.

Phosphorus

Inorganic and organic phosphorus compounds also have been found to be effective fire retardants in many polymers and are used as extensively as the halogens. Some phosphorus fire retardant additives are:

Tricresyl phosphate
Tris (2,3-dibromopropyl) phosphate
Triphenyl phosphate
Trioctyl phosphate
Ammonium phosphate
Tris (2,3-dichloropropyl) phosphate
Poly-β-chloroethyl
triphosphonate mixture

Fire-retardant phosphorus reactive monomers are:

Tetrakis (hydroxymethyl) phosphonium chloride (THPC)
Diethyl N,N-bis(2-hydroxyethyl)aminomethyl phosphonate
Hydroxyalkyl esters of phosphorus acids

Various boron compounds have found some use as fire retardants, especially in cellulose. They are especially useful in suppressing the afterglow that often results from the use of halogen-antimony combinations in cellulose and other oxygen containing polymers. The most commonly used boron fire retardant is zinc borate.

Hydrated aluminum oxide has become increasingly important in recent years as a fire retardant additive.

2.5.3 Thermoplastic Resins

Included under this heading are polyolefins, styrene polymers, poly(vinyl chloride), acrylics, acetals, cellulosics, polyamides, and thermoplastics polyesters. The "big three" (polyolefins, poly(vinyl chloride), and styrene polymers) represent over 70 percent of the total plastics production.

As their name denotes, thermoplastics soften when heated. In a fire, such materials can soften enough to flow under their own weight and drip or run. The extent of dripping or running depends on thermal environment, polymer structure, molecular weight, presence of additives or fillers, etc.

Polyolefins

The major polyolefin plastics are low density polyethylene, high density polyethylene, and polypropylene.

Polyethylene is crystalline and its density is directly related to its degree of crystallinity. Polypropylene is isotactic and crystalline. Chemically, polyolefins are very similar to paraffin wax and they burn in somewhat the same way. They ignite easily, burn with a smoky flame, and melt as they burn.

Considerable effort has been devoted to decreasing the flammability of polyolefins. The use of additive systems based on combinations of halogen compounds and antimony oxide has been the most effective. All of these fire retardant composition, however, can burn readily in a fully developed fire, contributing to the fuel load and producing very hot fires.

In their current major applications polyolefins present little fire hazard, but increasingly larger amounts are being used and many diverse applications are being developed. This is accentuated by the increasing use of polyolefin-based structural foams and moldings as well as their applications in relatively large items such as pallets, furniture, and bins. Therefore, danger exists that excessive fuel loads and paths for rapid flame spread may be created in the future.

Chlorinated Polyethylene

Polyethylene can be chlorinated in the presence of light or free radical catalysts to give a chlorinated polymer, the chlorine content of which can vary considerably depending upon the extent of chlorination and reaction conditions. Flammability decreases directly with the chlorine content. Various compositions containing 25 to 40 percent chlorine by weight are reported to extinguish under ASTM D-635 conditions. Compositions containing as much as 67 percent chlorine have been prepared. As with chlorinated polymers in general, antimony oxide enhances the efficiency of the halogen and, within limits, reduces the amount of chlorine required to yield the desired fire retardant properties.

Styrene Polymers

Polystyrene is an amorphous linear polymer with the formula:

$$\left[CH_2 - \underset{\displaystyle C_6H_5}{CH} \right]_x$$

Styrene polymers burn readily with the evolution of dense smoke. Depending on composition, molecular weight and conditions of burning, some materials drip flaming molten polymer. Fire hazard can be reduced by the use of additives and variations in composition that decrease ease of ignition and reduce the rate of flame spread under conditions involving relatively small ignition sources and/or low energy environments. In intense fires, even the best of these compositions burn rapidly. Pound for pound styrene polymers represent about twice the fuel load of cellulose. Foam products, on a volume basis, have relatively low fuel values, but can have high burning rates.

In the major current applications, styrene plastics present little fire hazard. However, with the increasing amount and diverse uses of polystyrene foam, in addition to the proliferation of large thermoformed items potential hazards arise when relatively large amounts of polymer are stored and when large surface areas are exposed. The hazard is enhanced by the high burning rate of polystyrene and by the high temperature and dense smoke generated in polystyrene fires.

An extensive array of additives has been tried to reduce the flammability of styrene-based products. The majority of these additives are halogen compounds and usually incorporate a synergist such as antimony oxide. Some contain phosphorus with or without halogen. The systems are considered to function by disrupting the free radical chain process of combustion in the gas phase. Other additives are based on a peroxide or on other free radical precursors. These are believed to function by increasing the depolymerization rate and promoting dripping of molten polymer, thus removing heat and flame from the burning sample.

Polyvinyl Chloride (PVC)

Because of its low cost and useful properties, PVC is one of the largest volume commercial thermoplastics. In 1974, more than two million metric tons were produced for a wide variety of applications.

PVC itself does not burn under most normal conditions; however, it generally is compounded with significant and often large amounts of plasticizers or processing aids, many of which are flammable (particularly the widely used phthalate, sebacate, and adipate esters, and various low molecular weight adipate polyesters).

When exposed to flame or to excessive heat, PVC emits hydrogen chloride at relatively low temperatures in a highly endothermic process. This emission together with the fact that the polymer contains more than 50 percent chlorine by weight, accounts for the low flammability of the uncompounded polymer. Depending upon the amount or type of compounding ingredients used during fabrication, decomposition products may include benzene, hydrocarbons, char and other fragments.

The most commonly used additive is antimony oxide because it acts as a synergist with the chlorine contained in the polymer. As little as 2 parts of antimony trioxide per 100 parts of resin (phr) give excellent fire retardance to compositions containing up to 50 phr of primary plasticizers such as dioctyl phthalate.

Chlorinated or phosphorus-containing plasticizers also are used in large quantities to reduce the flammability of plasticized compositions.

Phosphates, particuarly tricresyl phosphate, cresyl diphenyl phosphate, and 2-ethylhexyl diphenyl phosphate, have traditionally been added to PVC as plasticizers. They also enhance fire retardance and achieve excellent flame-out times.

Other widely used fire retardant PVC plasticizers are chlorinated paraffins. These relatively inexpensive materials contain about 40 to 60 percent chlorine and are often used in combination with phosphate plasticizers to reduce cost and to obtain special combinations of properties. They can also be combined with antimony oxide.

Vinylidene Chloride Polymers

Poly (vinylidene chloride) possesses an unusually low permeability to gases and vapors. This property can be exploited best in form of films or coatings and it is in these forms that vinylidene polymers find their greatest utility.

Most common test methods indicate that vinylidene chloride polymers have a low degree of flammability. They have a LOI of 60, which is higher than that of most thermoplastics produced in larger volume. This resistance to combustion is not only a function of their high chlorine content, but also is related to their tendency to dehydrochlorinate into a carbonaceous char highly resistant to combustion. Large volumes of hydrogen chloride are produced as a byproduct of this carbonization reaction.

Acrylics

Acrylics are polymers formed from acrylic or methacrylic esters.

The major plastic in this group is essentially a homopolymer of methyl methacrylate, a crystal clear material that softens at about 100°C. Various other acrylic polymers and copolymers are used in lacquer, enamel, and latex coatings.

Poly (methyl methacrylate) (PMMA) ignites readily and softens as it burns; it undergoes pyrolysis from the heat of the ignition source, the heat of combustion, or other environmental energy. The volatile products of pyrolysis then burn in the gas phase. Heat from combustion causes additional pyrolysis and consequent continued burning. PMMA is comparable in its flammability to the polyolefins. Its depolymerization energy (about 14 kcal) is low compared with that of polystyrene or polyethylene (about 20 kcal). Consequently, the pyrolysate contains much monomer and few high molecular weight products. Depolymerization rather than chain scission causes PMMA to drip less during burning than polystyrene or the polyolefins. Depolymerization together with the presence of oxygen in the ester group, probably accounts for PMMA's sparser smoke formation per unit weight in comparison with that of hydrocarbon polymers under most burning conditions. Reduced smoke formation probably also contributes to PMMA's experimentally established lower rate of burning because the less luminous flame feeds back less radiant energy to cause further pyrolysis.

Halogen and antimony compounds have been used to reduce burning rates and ease of ignition of PMMA, but less effort has been devoted to the fire retardation of PMMA than to that of other polymers. This lack of effort is due partly to the realization that it is difficult to inhibit the "unzipping" depolymerization mechanism so characteristic of this polymer. In addition, excellent transparency and aging characteristics are prime properties of most applications of PMMA; fire retarding additives usually detract from one or both of these properties.

Nylons

The major nylon plastics are nylon 66, and nylon 6.

Other molecular structures such as nylon 610, 612, nylon 11, and nylon 12 with different chain legnths between the amide groups are used for special purposes. Nylon polymers owe their wide application as engineering plastics to their strength, toughness, and solvent resistance.

Nylon moldings are often described as "self-extinguishing" due largely to their tendency to drip when ignited. If dripping is prevented, nylons burn with a smoky flame; they pyrolyze to a complex mixture of hydrocarbons, cyclic ketones, esters, and nitriles with some carbonization.

Fire retarding formulations are generally based on phosphorus-containing, or halogen-containing additives, with or without the addition of antimony; iron oxides and hydrated alumina. None of these systems prevents burning in a fully developed fire.

Many molding and extrusion compositions contain glass fibers or particulate mineral fillers (as much as 40 percent by weight) to enhance certain engineering properties. Such filled materials can burn more readily than the unfilled counterparts because the fillers tend to reduce dripping. Highly filled materials, however, have lower fuel value.

Currently nylons are used in relatively small items and have not, in themselves, posed serious fire safety problems. However, since larger products (e.g., gas tanks, large castings, automotive body parts, other structural applications, etc.) are being considered, more attention needs to be given to the analysis of the fire hazards that might be introduced.

Cellulosics

Regenerated cellulose and cellulose acetate are widely used as fibers. Cellulose derivatives that are useful as plastics include regenerated cellulose, organic and inorganic esters of cellulose, and cellulose ethers.

Regenerated cellulose (viscose) film is called cellophane. If cellophane is to be used as a moisture barrier, a very thin film of wax, nitrocellulose poly (vinylidene chloride), or other appropriate material is applied as a coating.

Cellulose nitrate plasticized with camphor (celluloid), the first thermoplastic, was introduced in 1870, but today it accounts for only a few percent of the cellulosics market.

Cellulose acetate (plasticized) is tough, clear injection moldable plastic with good electrical properties, but it has a low softening point and high water adsorption. The higher esters, propionate and butyrate, have lower water adsorption and process easily. Because they are hard, tough, strong, and transparent, they are used in film and sheet form, in injection moldings, and in the extrusion.

Ethyl cellulose is the most widely used cellulose ether. It finds application in molding, extrusion, and sheet fabrication. It is transparent and tough even at low temperatures. Typical manufactured items are flashlight casings and electrical appliance parts.

The basic flammability characteristics of regenerated cellulose are comparable to those of cotton.

Cellulose nitrate ignites readily and burns vigorously, generating nitric oxide gases. It presents the greatest fire hazard of all plastic materials, and thus, the use of nitrocellulose lacquers in some applications could be extremely hazardous from the point of view of fire spread. Temperatures as low as 150°C can initiate decomposition and ignition.

Organic cellulose esters and ethers melt and drip as they burn and give off a yellow sooty smoke. Halogen and phosphorus-containing plasticizers have been used to produce fire retardant grades, but they still burn readily in a fully developed fire.

Polyacetals

The commercial polyacetals are formaldehyde polymers and copolymers terminated ("capped") with esters or other groups for stabilization. Polyacetals are widely used for their mechanical characteristics; they are tough, solvent resistant, have low friction coefficients, and can be fabricated easily by injection molding. Major applications are automotive parts, appliances, and plumbing fixtures.

Formaldehyde polymers and copolymers are characterized by low oxygen index (high flammability), very low smoke, and low fuel value.

Little success has been achieved in fire retarding polyacetals because of the nature of their pyrolysis, which resembles that of PMMA. On the other hand, many of these polymers are used in relatively small parts where they do not present major fire hazards.

Polyesters

The polyesters included here are linear thermoplastic poly(ethylene terephthalate) (PET) and poly (butylene terephalate) (PBT). PET, chemically identical with polyester fibers and Mylar film, never developed a large market as a plastic due to difficulties in injection molding. However, PBT overcame these limitations and is a rapidly growing engineering thermoplastic. The current major applications are in small automotive parts and in the electrical/electronic industry.

These polymers burn with a smoky flame accompanied by melting, dripping and little char formation. A number of fire retarded grades are available; they are generally prepared by incorporating halogen-containing materials as part of the polymer molecules or as additives. Metal oxide synergists are frequently included. These fire retarded systems are resistant to small ignition sources in low heat flux environments, but still burn readily in fully developed fires.

As with other engineering thermoplastics, many end uses are in small parts that may not constitute a significant fire safety hazard in themselves. Care must be exercised in electrical and electronic applications to avoid possible ignition from electrical sources.

Polycarbonates

Polycarbonates are a special class of polyesters derived from bisphenols and phosgene; some contain fibrous or particulate fillers; they are extremely tough and have good creep resistance. Over 40 percent of polycarbonate production is used in appliances and electronic devices; other uses are in transportation, sports equipment, signs and glazing.

Commercial unmodified polycarbonates are significantly less flammable than unmodified styrene, olefin, or acrylic polymers. Some char is produced during pyrolysis or burning. They extinguish during simple horizontal burning tests and have an oxygen index significantly above those of all the previously discussed unmodified thermoplastics. Their fire resistance has been further improved by the use of halogenated bisphenols in the preparation of the polymer or by the use of halogen-containing additives with or without antimony oxide.

A number of copolymers containing high percentages of acrylonitrile or methacrylonitrile have recently been developed specifically for packaging carbonated beverages and food items. The common characteristics of these materials are their low transmission of oxygen, carbon dioxide, and water. No details are available on the flammability of these polymers. They have been reported to be disposable in normal waste channels without adversely affecting incineration operations or causing changes in effluent compositions.

Poly(aryl ethers)

PPO, a commercially available poly(aryl ether), is prepared by the oxidative coupling of 2,6-xylenol to give:

$$\left[-C_6H_2(CH_3)_2-O- \right]_x$$

This material is a rigid, tough, chemically and thermally resistant thermoplastic with good electrical properties. It has been proposed for various engineering and electrical applications. Glass-filled modifications are available. Major applications are in automotive and electrical parts.

Poly(aryl ethers) are char forming; blend with polystyrene (Noryl), reduce the dripping of polystyrene and make ignition more difficult. Fire retarded versions, based on halogenated coreactants or organo-phosphate additives are available.

2.5.4 Thermosetting Resins

Polymerization of thermoset resins generally is divided into two stages: in the first stage, a relatively low molecular weight prepolymer is formed which can be melted, dissolved, or molded; in the second stage, the prepolymer is crosslinked (cured) with or without the application of heat in the presence of suitable catalysts, activators or promotors.

Thermosetting resins are produced in large quantities and are extensively used in construction, housing, and large appliance industries where they may contribute significantly to the fire load in a particular area or product. Consequently, their fire safety characteristics are of primary concern. Because of their crosslinked nature, thermosets generally do not soften or drip when exposed to a flame as do many thermoplastic materials. Their flammability is a function of the thermal stability of the primary chemical bonds and the ease with which volatile gaseous products can be produced by pyrolytic processes to provide fuel for a self-sustaining fire. Many thermosets (e.g., the phenolic resins) produce very little flammable fuel when heated by an ignition source. They produce an insulating char that can only be oxidized at extremely high temperatures or high oxygen concentrations. Burning of such materials can be a slow process under many conditions since the polymer substrate is protected by the surface char. Such resins are inherently fire retardant and will pass many common laboratory tests without the need of a fire-retardant modification or additive.

Phenolic Resins

Phenolic resins are prepared by the condensation of phenol and formaldehyde.

Physical properties of phenolic resins vary widely depending upon the type, kind and amount of filler, kind of reinforcement, ratio of phenol to formaldehyde, type of curing catalyst, and other formulation variables.

Cured phenolic resins do not ignite easily because of their high thermal stability and high charring tendency in the presence of fire and heat. The principal volatile decomposition products are methane, acetone, carbon monoxide, propanol, and propane. A variety of additives, in addition to increased crosslink density, have been found to be useful in applications where a degree of fire retardance above that inherent in the usual polymer is required. The following fire retardants are most commonly added:

1. Halogenated paraffins
2. Phosphorus compounds and halogenated phosphorus compounds
3. Hydrated alumina or zinc borate
4. Synergists.

Unsaturated Polyester Resins

Two classes of thermoset resins are commonly referred to as polyester resins — the unsaturated polyester resins and the alkyds. Unsaturated resins are prepared by condensing a saturated dibasic alcohol plus both a saturated and an unsaturated dicarboxylic acid into a prepolymer which is then dissolved in a vinyl monomer, usually styrene. The cured resin is produced by free radical copolymerization of styrene monomers and unsaturated acid residues.

The resins usually are compounded with a reinforcing glass fiber or filler before curing.

A very useful fire retarding filler for polyester resins is hydrated alumina. Although the degree of fire retardance obtainable by the sole use of this additive appears to be considerably less than that which can be obtained by the use of a halogenated resin, the structural reinforcing property combined with the relatively low cost of the additive indicates advantageous combination with more efficient fire-retarding methods. In addition, hydrated alumina assists in reducing the smoke generated by the polyester composition and does not yield corrosive hydrogen halide when subjected to fire conditions. This combination of low cost, low smoke, structural reinforcement, and reduced corrosivity has led to widespread use of this filler in applications where maximum fire retardance is not required.

Phosphorus compounds and chlorinated waxes have been the most common organic fire retarding additives for polyester resins. Chlorinated biphenyls and organic antimony compounds also have been used to some extent, but the former are not known to pose a health hazard. High-melting insoluble chlorocarbons, such as bi(chlorendo) cyclooctane have recently come into use because of their high resistance to leaching, good thermal stability, and good fire retardance efficiency.

Modification of the saturated acid component has been by far the most successful commercial method of preparing fire retarded unsaturated polyester resins. Of the commercial fire retarded unsaturated polyesters currently being marketed, the majority incorporate chlorendic acid or its anhydride. Other halogenated dibasic acids used in these polyesters are tetrachlorophthalic and tetrabromophthalic acids. The greater fire retardance of tetrabromophthalic anhydride permits the preparation of unsaturated polyester resins with excellent fire retardance at bromine contents as low as 15 percent by weight.

Another method of incorporating fire retardance into a polyester resin involves the addition of suitable halogen- or phosphorus-containing crosslinking agents. Although various degrees of fire retardance can be obtained using this method, the cost of the modified unsaturated compounds generally precludes their use in sufficient quantities to obtain satisfactory fire retardance by this means alone. The crosslinking agents are used to a limited extent in combination with other types of fire retardants to obtain special properties unattainable by other means.

Both fire retarded and conventional unsaturated polyester resin formulations yield copious amounts of smoke when exposed to a developed fire because styrene is the major product of pyrolytic decomposition and styrene burns with a very smoky flame. High smoke values have only been marginally reduced to date by the use of relatively large amounts of inorganic fillers such as alumina hydrate. Although considerable research is currently being carried out industrially to overcome this disadvantage, only minor improvements have been achieved.

Alkyd Resins

Alkyd resins are also unsatured polyesters, but differ considerably in their structure from styrenated polyesters. The first stage polymer is a condensation product of a dibasic acid (usually phthalic acid) and a dihydric alcohol (typically ethylene glycol). The unsaturated carbon-carbon bonds, however, do not result from copolymerization with an unsaturated dibasic acid or alcohol, but from incorporation of vegetable oils such as linseed oil, soybean oil, or tung oil. These oils are triglycerides containing unsaturated carbon-carbon bonds in the fatty acid components. The oil is usually reacted directly with glycol to form a monoglyceride, which is then reacted with acid to form alkyd resin. Thus, in these resins, the unsaturation occurs not in the main chain of the polyester but in its side chains. Crosslinking (hardening) to form the second stage resin is accomplished not by copolymerization with styrene but by air oxidation of the unsaturated groups. Oxidation is usually aided by the addition of certain "driers" such as cobalt naphthenate and lead soaps.

Alkyd coatings are of value because of their comparatively low cost, durability, flexibility, gloss retention, and reasonable heat resistance. They may be modified with rosin, phenolic resins, epoxy resins, and monomers such as styrene.

Because of their wide use as coatings, the fire safety characteristics of alkyds are generally modified either by formulating them as intumescent coatings or by incorporating the conventional halogen-antimony oxide fire-retardants. Of these latter, the two most generally used halogen-containing additives are chlorinated waxes, or chlorendic acid or some other halogenated carboxylic acid incorporated in the polyester molecule.

Antimony compounds, generally antimony oxide, are added as one of the pigment filler ingredients in the final formulation of the coating. This approach can lead to either an intumescent or an nonintumescent coating.

Epoxy Resins

Epoxy resins are specialty thermosetting resins that have developed a significant market as adhesives and coatings, in reinforced laminates, and in molding and casting applications because of a unique combination of properties. The most important of these properties are excellent adhesion, corrosion resistance, toughness, and abrasion resistance.

Epoxy resins generally are prepared by reacting a first stage poly-functional epoxy compound or resin with a basic or acidic crosslinker (or "hardener") to yield a thermoset product crosslinked by ether linkages.

Although epoxy resins normally are flammable, their flammability can be reduced considerably by the use of a variety of phosphorus- or halogen-containing additives or reactive monomers. Tris (2,3-dibromopropyl) phosphate is one of the most common fire retarding additives, and tetrabromobisphenol A is by far the most widely used reactive fire retardant monomer. The need for fire retardance in epoxy resins has been relatively small. Consequently, a relatively small amount of fire retardant epoxy resins except for printed wiring boards is sold annually.

2.5.5 Specialty Plastics

The materials in this group fall into three general categories: 1) aromatic and heteroaromatic polymers that are generally used for their high-temperature resistance; 2) fluoropolymers that are generally used for their resistance to temperature, chemicals, and combustion; and 3) chlorinated poly-ethers.

Polyimides and Poly (imide-amides)

Aromatic polyimides and poly (imide-amides) and their modifications exist in a variety of chemical structures and physical forms. They find applications as injection moldings, compression moldings, laminates, films, foams, and coatings. Their outstanding features are oxidative stability and dimensional stability at elevated temperatures for extended periods of time.

Aromatic polyimides are characterized by high char formation on pyrolysis, low flammability, and low smoke production when immersed in a flame; the char yield is about 60 percent of the initial weight. Aromatic poly-ether-sulfones are strong, tough, rigid engineering thermoplastics with good thermal stability and creep resistance; they are finding applications in the electrical, automotive, and aircraft industries where special combinations of mechanical, thermal or fire resistance properties are desired.

Aromatic polyethersulfones, along with some newer polyimides and poly (phenylene sulfides), are among the most fire resistant non-halogenated thermoplastics. These materials are also very difficult to process. Their fire retardance is probably due to their char-forming tendencies.

Polyphenylenesulfide is a high melting (285°C) crystalline material of outstanding chemical resistance, thermal stability and fire resistance. It is available in grades suitable for coating, injection molding, and compression molding. Rigidity is retained up to 250°C and the polymer is unaffected by long exposure at 230°C. Major uses are as chemical-resistant coatings for use at elevated temperatures, as molded parts for pumps, valves, impellers, seals, etc., and as electrical parts.

Poly (phenylene sulfide) is a char former. In an inert atmosphere essentially no volatiles are evolved below 500°C and 40 percent char remains at 1,000°C. Fire resistance is comparable to that of the best polyethersulfones. It is probably the least flammable of any of the nonhalogen containing thermoplastics.

An aromatic polyester prepared from p-hydroxybenzoic acid was introduced in 1970 under the name of Ekonol. It has been shown to be thermally stable in air for extended periods at 315°C. Because of the difficult fabrication techniques required, its use is limited to such relatively small scale applications as electrical insulators and circuit boards, self-lubricating bearings, and slip free coatings for household cooking utensils and automobile piston rings. The limitation of the polymer to small scale applications and metal coatings does not make its fire resistance especially important commercially.

Poly (vinyl fluoride) is sold only in film form (Tedlar film). It is widely accepted for surfacing and also industrial, architectural, and decorative building materials. Its attributes are outstanding weatherability; inertness to solvents, chemicals and stains; as well as excellent abrasion resistance and cleanability. Current applications include surfacing film for wall panels, vinyl fabrics, aluminum siding, hard board panels for siding, and aircraft and automotive interiors.

Although chemically quite similar to poly (vinyl chloride), poly (vinly fluoride) is much more flammable. Poly (vinyl fluoride), like PVC, appears to pyrolyze in two stages. There is no weight loss up to 250°C. Hydrogen fluoride begins to evolve at about 250°C. At higher temperature, considerable chain scission occurs with formation of volatiles, which form fuel for the fire. Hydrogen fluoride is much less efficient in quenching as phase combustion than hydrogen chloride or hydrogen bromide.

Poly (vinylidene fluoride) is fabricated by molding, extrusion, and coating. Its outstanding combination of physical, electrical, thermal, and chemical properties account for its major markets in electrical insulation, chemical processing equipment and long-life finishes for metal siding. It is widely used in the electronics industry and as jacketing on aircraft wire.

Poly (vinylidene fluoride) is thermally more stable than poly (vinyl fluoride), but not so stable as polytetrafluoroethylene. Rapid pyrolysis occurs at about 425°C with evolution of hydrogen fluoride and formation of a char that amounts to about 40 percent of the initial weight.

Polytetrafluoroethylene (TFE) and fluorinated ethylenepropylene (FEP) copolymer, are composed wholly of fluorine and carbon. The major difference in the two polymers is the lower melting and easier processing of FEP. These polymers are dense, chemically inert, thermally stable, and have low coefficients of friction. The applications, based on these properties, are in the chemical, electrical, and mechanical areas. Electrical uses include aircraft and other wiring, and molded insulators. Use is bases, hydraulic systems, valves, seals, linings, and gaskets take advantage of the chemical inertness of these polymers.

TFE and FEP are certainly among the least flammable of polymers, and their LOI is about 95. They do not support combustion. At about 500°C, TFE pyrolyzes almost completely to the fluoroolefin monomers. These monomers can oxidize under certain conditions to carbonyl fluoride a corrosive toxic gas.

Chlorotrifluoroethylene polymers are similar in many respects to the completely fluorinated materials. They are generally more rigid and expensive, and somewhat less thermally stable than TFE or FEP polymers. Their applications are based on chemical resistance, low water vapor transmission, good electrical and mechanical properties, and processability. Uses include insulation for hook-up wire and cable jacketing, process valves, fittings, seals, tubing film, containers and closures for corrosive materials, and packaging medical supplies.

CTFE polymers, just as their completely fluorinated analogs, are very resistant to burning.

The 1:1 ethylene copolymer forms a friable char on exposure to flame and does not melt and drip away from the flame. It has a LOI of 60.

2.5.6 New Materials Under Development

The following summary identifies some of the more fundamentally different new materials that offer potential for improved fire safety. These materials are presently under various stages of development and study.

Heteroaromatic Polymers

Since 1961, considerable effort has been spent on the synthesis of polymers that retain their properties at high temperatures. The impetus for this work arose from the synthesis of polybenzimidazole by Vogel and Marvel (1961) and the work on polyimides by Sroog, et al. (1964). These materials and the follow-up structures were characterized by being composed primarily of aromatic and heteroaromatic rings. This structure generally leads to high char formation and flow flammability.

Polybenzimidazole (PBI), has been extensively studied, particularly for fibers, but also for matrix resins, adhesives, coatings, and foams. It has excellent fire resistance in all physical forms. A major problem with use of PBI for plastic applications is the difficulty in fabrication. This is generally true of all high-char-forming heteroaromatic polymers. They are usually fabricated from a prepolymer, the conversion to final polymer taking place in the mold in much the same way in which thermosetting resins are handled.

Many other heteroaromatic structures have been studied. Among these are:

polyquinoxalines

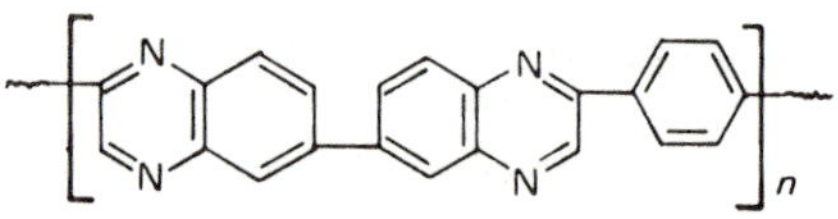

polyoxadiazoles

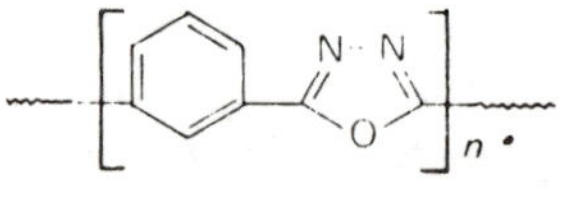

polybenzothiazoles

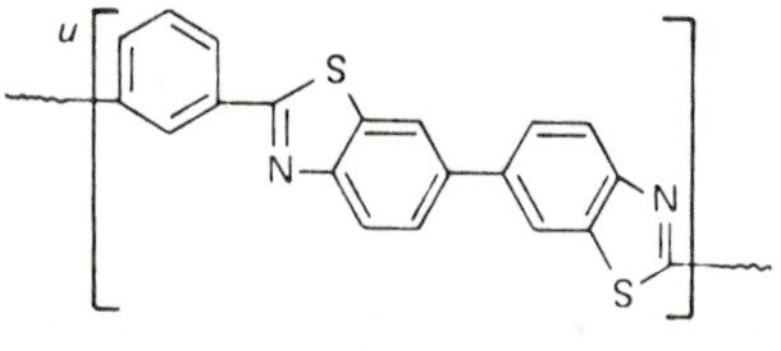

and BBB, pyrrones, and related "ladder" polymers

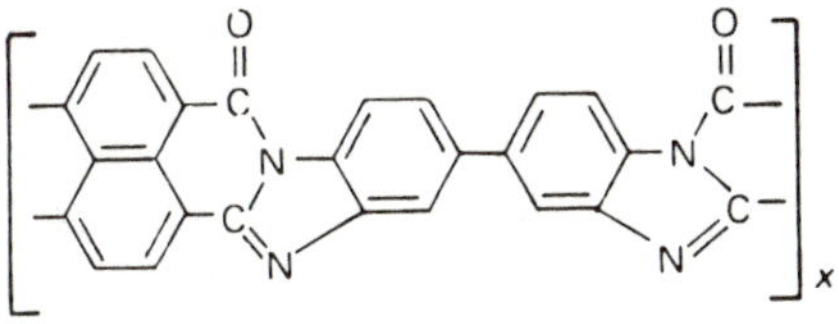

All of these suffer from relatively high cost of raw materials, polymer preparation, and fabrication.

Polyphenylene represents another class of high-char-forming resins that have good resistance to burning. Cost and/or relatively expensive processing methods have been limitations.

Recently, a new class of thermosetting polyphenylenes, the H-Resins, were announced. These materials are reported to be readily curable and useful as filled, reinforced, or unfilled materials. Their LOI is stated to be greater than 55.

A variety of aromatic polyesters with potentially improved fire safety characteristics have been reported in various stages of development. The improvements in the fire safety characteristics of these materials are based on increased char formation.

Morgan has synthesized a large number of aromatic polyesters with large cross-planar and large three-dimensional substituents. Many of these polyesters have high softening points and produce large amounts of char under pyrolysis. One of these, the polycarbonate from phenolphthalein has been found to be potentially useful as a fire resistant glazing material for aircraft.

Teijin, Ltd. (Japan) has announced developmental quantities of poly(ethylene naphthenate) (PEN) in film form ("Q" film), which might be expected to have greater resistance to burning than poly(ethylene terephthalate). Unitika, Ltd. (Kyoto, Japan) has reported on a polyarylate based on iso- and terephthalate esters of bisphenol A. Its flammability is reported to be less than that of conventional polycarbonates.

Bis-Maleimides

This type of polymer was developed to overcome some of the fabrication deficiencies of the polyimides. The monomer, is thermally polymerized with evolution of little or no volatiles to give a very thermally resistant, high-char-forming, low flammability product (Mallet and Darmony, 1974). Bis-maleimides have been investigated primarily as matrix resins for composites (Gilwee et al., 1973).

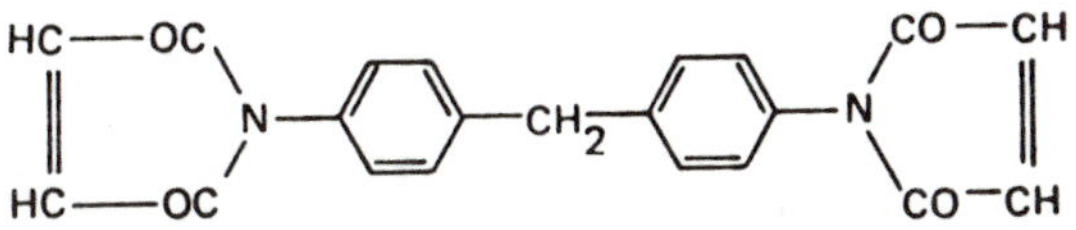

Composites and Laminates

An enormous quantity of composites and laminates, made from various synthetic plastics, wood products, various fillers, reinforcing agents, and adhesives, is produced annually in the United States. Like fiber blends, composites and laminates may pose special fire hazards. In many cases, the exact composition of these materials is proprietary. In any case, however, their fire performance cannot be predicted from a knowledge of the performance of the components. The unsaturated polyesters, phenolics, epoxies, and amino resins, are used most often in the manufacture of composites and laminates.

2.6 Foams

2.6.1 Introduction

Synthetic foams can be made from practically any polymer. Foamed natural and synthetic polymers pose special fire hazards. Their fire safety aspects are considered here.

Flexible foams find application mainly in cushion, padding, mattresses, and carpet or rug underlays.

Although some *rigid foams* are used in sandwich constructions for aircraft, furniture and building structures, the major application of rigid foams lies in the field of thermal insulation. Because of their versatility and low cost, this use of synthetic foams is growing rapidly.

Rigid closed-cell foams are good thermal insulators, whereas the flexible foams which are open-cell are not good thermal insulators.

2.6.2 Fire Safety Aspects of Foams

The rate of pyrolysis under the impact of a given heat flux, and the heat generated during the subsequent combustion of the volatile pyrolytic gases, are perhaps the most important variables in determining the fire safety characteristics of polymeric solids. Since polymer compositions burn only on their surface, the amount of surface area available for combustion is important in determining the rate of combustion and, therefore, the intensity of the flame. The high surface area per unit weight of material being subjected to pyrolytic conditions necessarily increases the flammability of a foam beyond that of the polymer composition from which it is made. The most important difference is the density and the highly insulative properties of the material. The density of most commercial plastics is in the range of 56–75 lb/ft^3 (0.9–1.2 g/cm^3). Since many foams have densities of about 2 lb/ft^3 (0.032 g/cm^3), only about 2.5 to 3.5 percent of the total volume of such a foam is composed of solid polymer. Incorporation of such a high volume of gas into the polymeric structure affects the burning characteristics of the material in several ways. First, since a greater surface is exposed to the oxygen of the air, the rate of pyrolysis and burning is increased. Additionally, the high gas content gives foams a low specific heat per unit volume. Their low thermal conductivity tends to concentrate the heat on the surface of the structure rather than dissipating it to underlying material or substrate. The result is a rapid temperature rise and pyrolysis of the surface material when exposed to a flame. This often leads to an extremely rapid flame spread rate. However, other factors may moderate this effect considerably. For example, the small amount of potentially flammable material per unit volume in low-density foams results in a very small amount of total heat being available per unit area for flame propagation.

Thus, if the foam is a thermoplastic (e.g., polystyrene) the heat of a flame rapidly melts the foam adjacent to it and the material may recede so fast from the flame front that there is no ignition. A highly crosslinked thermoset foam, on the other hand, behaves in an entirely different manner. Since little or no melting occurs, the surface does not recede from the flame front and the foam is immediately ignited. The flame then spreads if the foam is flammable. However, a fire retarded foam, under the same conditions, can pyrolyze rapidly in the vicinity of the flame and leave a carbonaceous char on the surface of the material. This highly insulating char protects the remainder of the material from the effects of the flame.

2.6.3 Rigid Foams

Like plastics in general, rigid foams may be subdivided into thermoplastic and thermosetting foams. Thermoplastic rigid foams can be prepared from most thermoplastics (e.g., polystyrene, ABS resins, polyethylene, polycarbonate, poly (vinyl chloride)); they generally melt or depolymerize in a fire (i.e., fragment into small, usually volatile molecules). Some thermosetting foams, such as the phenolics, also characterized as semi-rigid crosslinked foams, do not melt or depolymerize, but, for the most part, char in place instead. On the other hand, crosslinked polyurethane foams may depolymerize, melt, and drip like some thermoplastics.

Rigid foams may be fire retarded in several ways; including the use of fire-retarded additives such as organic phosphate derivatives either in the foam formulation or in a coating applied to the foamed surface. Both techniques effectively reduce the burning rate of the foam surface. A more recent practice is the introduction of an inert inorganic filler to control flame spread and flashover.

An alternate method of achieving fire retardation in foams consists of making them from fire retardant polymers. In one approach, halogens such as chlorine or bromine are introduced into the backbone of the basic polymer. Foams generated from bromoepoxy prepolymers and chlorendic anhydride as curing agent are examples.

Another approach consists of using high-char-yield materials such as the polyisocyanurates, polyimides, polybenzimidazoles, polyquinoxalines, and polyphenylenes. These polymers directly yield fire retardant foams capable of passing current flammability tests without the need for addition of fire suppressant additives.

Rigid Polyurethane Foams

Rigid polyurethane foam and expanded polystyrene foam are the volume leaders in the thermosetting and thermoplastic classes, respectively.

Polyurethanes are the reaction products of a dihydroxylic or polyhydroxylic compound or resin and a diisocyanate or polyisocyanate. Polyurethane foams are prepared by the controlled introduction of a gas phase during the fundamental reaction so that the permanent cellular structure is produced.

The cellular nature of polyurethane foams generally influences their flammability. A high surface temperature is generated by an ignition source because of low thermal conductivity. This effect can cause almost instantaneous conversion to flammable gases resulting in high surface flame spread and high flaming temperatures once the surface is ignited. Crosslink density significantly affects the flammability of polyurethane foams.

In general, fire retardance is imparted to polyurethane foams by the chemical incorporation of halogen and/or phosphorus compounds into the material. Chemical modification of the polyol with phosphorus or a phosphorus-chlorine combination is currently employed in the commercial preparation of fire retardant foams.

The use of phosphorus in fire retardant polyurethane foams leads to high char formation combined with easy processing because of the relatively low viscosity of most phosphorus compounds. This combination of desirable properties has made phosphorus compounds, with or without halogen, the most widely used fire retardants in polyurethane technology. Reactive phosphorus compounds, such as Fyrol 6 (Stauffer Chemical Company) are used extensively. They are added directly to the polyol. Fryol 6 has enjoyed considerable commercial success and currently appears to be the most widely used fire retardant for polyurethane foams.

$$(CH_3CH_2O)_2\text{-}\underset{\underset{O}{\parallel}}{P}\text{-}CH_2N(CH_2CH_2OH)_2$$

Polyurethane foams may be fire retarded also by incorporating non-reactive additives that act as fillers or plasticizers. The most commonly used example of the latter is tris (2,3-dibromopropyl) phosphate. Non-reactive additives have not been used extensively because of their fugitive nature and their tendency to migrate from the foam under many conditions of extended use.

Although polyurethanes themselves are nontoxic, their pyrolysis products have been shown to contain considerable quantities of toxic gases. Significant amounts of hydrogen cyanide have been detected in polyurethane combustion products although the relative toxic hazard of these materials in gaseous mixtures containing large amounts of carbon monoxide has not been definitely established.

Petajan, et al. (1975) have recently reported the isolation and identification of a highly toxic phosphate ester in the combustion products of fire retardant urethane foams containing a combination of trimethylolpropane polyol and Fryol 6 as the active fire retardant. Since even the small concentrations observed were shown to be considerably more toxic to rates than the much larger concentrations of carbon monoxide, the problem of toxicity of fire retarded polyurethane combustion products is now under extensive investigation.

Polystyrene Foams

The use of polystyrene foam for 1974 has been estimated as 189,000 metric tons and is expected to increase to 445,000 metric tons by 1980; it accounts for 40 to 50 percent of the total rigid foam market. Packaging is the major end use, consuming about two-thirds of the total produced. Construction, appliance, and marine uses make up most of the remainder.

Polystyrene foams can be classified according to density ranges and by processing methods. Expanded bead foams are usually in the density range of 1 to 2 lb/ft^3 (0.016 to 0.032 g/cm^3) and account for approximately two-thirds of the polystyrene foam used today. The material for bead foaming consists of beads of particles of polystyrene with 6 to 7 percent pentane imbibed into the beads during or after polymerization.

Polystyrene has a high rate of combustion and burns rapidly with production of dense smoke. Other than carbon monoxide, common to the combustion of all organic materials, there are no known highly toxic combustion products associated with polystyrene.

Fire retarded versions of foamed polystyrene are more difficult to ignite with small ignition sources, but burn rapidly with high flame spread rates in the high energy environment of an intense fire. The blue-colored fire retarded foam for construction uses is reported to contain acetylene tetrabromide and copper phthalocyanine. Another way to reduce the fire hazard of polystyrene foam involves incorporation of a very small amount of additive that promotes molecular weight degradation, softening, and dripping when the foam is heated. Thus, the foam retracts from the ignition source and/or drips away carrying off heat flame.

In most current uses of foamed polystyrene there is little fire hazard. In packaging applications, the major hazards are in warehousing and storage of large quantities. Hazards in construction, appliance, and marine uses depend greatly on how and where the foam is used. These hazards vary from virtually none, when the material is used as insulation under a concrete slab on the ground, to severe when exposed and unprotected on interiors of building walls. The relative hazards in uses between these extremes, and the effectiveness of additives, chemical modifications, coatings, configuration, coverings, etc., under fire situations, need further study and definition.

A major concern is the projected growth and proliferation of structural foams applications. Significantly greater fire hazards may occur as concentrations of such materials increase in living, working, and storage areas.

Polyolefin Foams

Cellular polyolefin (polyethylene, polypropylene) plastics show better thermal, acoustical, and electrical insulating properties than their corresponding solid resins. They also have better dielectric properties as well as mechanical damping characteristics and are more flexible. Closed cell foams have densities up to 55 lb/ft^3 (0.88 g/cm^3).

Polyolefin foams are used mainly as protective cushion packaging for fragile merchandise. They also find applications in weatherstripping tapes, automotive body joint seal gaskets, backings for wood floorings, and flotation devices. The dielectric properties make foamed polyolefins with density of 30–40 lb/ft^3 (0.48–0.64 g/cm^3) very useful for telecommunciation cable insulation. Crosslinked polyolefin foams can be vacuum formed into items like helmet liners and automotive and aircraft safety padding.

Without fire retardant modifications, polyolefin foams burn cleanly at rates of 3 to 6 in./min in horizontal configurations. They have flame spread ratings (ASTM Test E84) of 10 to 20, depending on thickness.

Cushion packaging and gasket applications rarely, if ever, require fire retardation. Flotation devices (e.g., for aircraft seat cushions), invariably require fire retarding modifications.

Antimony oxide/chlorowax combinations or other aliphatic chlorine sources are generally used to fire retard polyolefin foams. Because of the higher processing temperature required (up to 280°C) when using azo blowing agents, the antimony oxide/chlorine system is inadequate and, in that case, phosphorus-containing fire retardant systems are preferred.

Poly (Vinyl Chloride) (PVC) Foams

The largest use for foam based on PVC is in coated fabrics, where the foam is sandwiched between a supporting fabric and a wear surface. This arrangement serves as an upholstery material used widely in transportation, and in home, business, and entertainment areas. It is also used as clothes linings and in accessories such as handbags, shoes, and boots.

Chemically blown film has also been used extensively in flooring. A foam lamina can be topcoated with a wear surface and placed onto a solid flooring substrate.

PVC is inherently fire retarded because of its high chlorine content. Flexible PVC foams present increased fire hazards because of the plasticizers they contain. Phosphate ester plasticizers should improve performance as to ignition and flame spread when compared to the more usual phthalate plasticizers.

Phenolic Foams

Phenolic foams are prepared by the condensation of phenol and formaldehyde (see Volume 1 for details). Although there has been considerable activity in the area of phenolic foams, these materials have not attained the prominence of the polyurethane foams because of their brittleness and open-celled structure.

Cured phenolic resins have good thermal stability and high tendency to char in an intense fire. Even after the removal of the ignition source, the foams often smolder and char until they are almost completely consumed. This latter phenomenon, called "punking," is claimed to be overcome by the use of boric acid/oxalic acid and ferric/aluminum chloride as the foaming catalysts. The addition of antimony compounds is also reported to decrease punking. Chlorinated aromatic compounds are claimed to achieve similar results when included in the phenolic foam formulation.

Urea-Formaldehyde (UF) Foams

UF foams are produced by several techniques including the use of hydrocarbons or fluorocarbon, *in situ* generation of carbon dioxide, and dispersion of air into the resin before acid curing of the foam. Foams produced by the hydrocarbon/fluorocarbon routes are generally closed-cell foams, whereas other techniques provide open-celled structures.

Whereas UF foams can be rated as materials which burn with difficulty, blending of UF with another polymer can decrease the resistance of the foam to burning. Fire retardant additives, e.g., phosphorus and boron compounds, decrease the flammability. UF foaming resins have been used in intumescent coatings for the protection of various substrates including wood.

Polyisocyanurate Foams

Isocyanurate-based foams have significantly increased in use during the past decade primarily due to the introduction of low-density foams and the mass production of skin-molded polyisocyanurates. The foams are prepared by trimerizing the appropriate isocyanates in the presence of suitable catalysts.

Integral skin foams represent a new opportunity for polyisocyanurate foams in the furniture industry as a replacement for polyurethanes foams. Their maximum service temperature is about 50°C higher than that of polyurethanes; their improved flame resistance, low fuel contribution, and low smoke generation make them attractive for applications where flammability is a serious consideration.

2.6.4 Flexible Foams

Flexible foams can be made from practically any elastomer. They are used in a variety of applications, the most important being mattresses, seat cushioning, rug underlay, and carpet backsizing.

When a chemical blowing agent is used in a dry-compounding recipe, the resulting foam rubber is generally referred to as sponge rubber. Sponge rubber is made mostly from natural and from styrenebutadiene rubber, although silicone and fluorocarbon (Viton) sponge rubbers are also available.

Approaches used for fire retardation generally are the same as those used with nonfoamed elastomers except that post-treatments similar, in principle, to those discussed in connection with wood and wood products are possible because of the cellular nature of foams. Thus rubber sponge has been impregnated with slurries of gypsum and cement. Polychloroprene foams can be made highly fire retardant by post-treatment with mixtures containing ammonium sulfamate or melamine-aldehyde condensation products.

Methods for fire retarding flexible polyurethane foams are essentially the same as those used with rigid polyurethane foams. Flexible foams, however, burn readily even when fire retarded; a totally satisfactory solution to the pressing problem of fire retardant flexible foam for cushionining has not yet been developed. The novel polymers possessing inherently improved fire safety characteristics, are all made from molecules which are too rigid to yield flexible foams. The phosphonitrilic elastomers are promising and should be evaluated.

2.7 Fire Retardant Coatings

2.7.1 Introduction

The use of fire retardant coatings is one of the oldest methods for protecting flammable substrates from reaching ignition temperatures and for preventing nonflammable substrates from reaching softening temperatures.

2.7.2 Nonintumescent Coatings

Although fire retardant nonintumescent coatings do not provide the same degree of fire protection to the substrate as to intumescent coatings, they are useful in marine applications, which constitute one of the largest areas of usage of such fire retardant coatings, because ships receive repeated paintings in an effort to provide corrosion protection. As layers of conventional paints build up, the paints pose a fire hazard regardless of the nature of the substrate. To combat this hazard fire retardant coatings were formulated so as to not sustain combustion, even when thick films were exposed to fire.

Alkyd Coatings

The most widely used fire retardant nonintumescent coatings are based on chlorinted alkyds. A common method for preparing fire retardant alkyl coatings is to use reactive chlorinated diacids or anhydrides such as: chlorendic anhydride or tetrachlorophthalic anhydride.

Halogenated additives, such as chlorinated paraffins, have been used extensively in fire retardant coatings because of their low cost and their minimal effect on regular properties when used in combination with metal oxides (antimony oxide is the most commonly used metal oxide).

Inert mineral fillers also reduce the flammability of coatings. They act simply by diluting the fire load presented by the combustible organic material; relatively high filler loadings must be used before any significant improvement is seen.

2.7.3 Conventional Intumescent Coatings

Intumescence is defined as "an enlarging, swelling or bubbling up;" it is well demonstrated by the high school chemistry experiment in which sugar is reacted with sulfuric acid. The acid dehydrates the sugar to a carbonaceous char, which is expanded into a foam by the steam produced by the heat of the reaction.

Intumescent coatings have been available commercially for over 20 years. They are used to prevent flammable substrates, such as wood and plastics, from reaching ignition temperatures, and nonflammable substrates, such as metals, from reaching softening or melting temperatures.

Conventional intumescent coatings contain several ingredients that are necessary to bring about intumescent action. An intumescent *catalyst* is used to trigger the first of several chemical reactions that occur in the coating film. The coating contains a *carbonific* compound that reacts with the intumescent catalyst to form a carbon residue, and a *spumific* compound that decomposes to produce large quantities of gas. The latter causes the carbonaceous char to foam into a protective layer. A resin binder forms a skin over the foam and prevents the escape of the trapped gases. Unfortunately, most (but not all) intumescent coatings have one or more of the following drawbacks: poor aging, poor weathering, poor humidity resistance, poor color, poor flexibility, and high cost. Virtually every common resin has been tried as a binder in the preparation of intumescent coatings.

Chlorinated resin binders, such as chlorinated rubber and vinyl polymers, are especially desirable because they also exhibit a spumific effect by the evolution of hydrogen chloride at elevated temperatures while simultaneously contributing significantly to char formation.

The main function of an intumescent *catlayst* is the dehydration of the carbonific compound to a carbonaceous char. Three inorganic acids – sulfuric, boric, and phosphoric – have been used; the latter being most common.

One of the important requirements of an intumescent catalyst is that it must decompose at a lower temperature than the decomposition temperature of the carbonific compound.

The *carbonific* compound is a polyhydric organic compound, which forms a carbonaceous char when it reacts with the acid produced by decomposition of the intumescent catalyst. It is this carbonaceous char that prevents or retards the flame spread. Monomeric carbonifics include sugars, polyhydroxylic compounds of the pentaerythritol family and other compounds. Among the polymeric carbonifics are urea/formadehyde resins, phenolics, pentaerythritol polyurethanes, and starch.

Spumific compounds decompose when heated, evolving large quantities of nonflammable gases such as ammonia, hydrogen chloride, and carbon dioxide. The residue contributes to the carbonaceous char. Commonly used spumifics are melamine, dicyandiamide, guanidine, urea, and chlorinated polymers.

For intumescence to occur with the above ingredients, five important and interrelated steps must occur in precise order as follows:

1. The intumescent catalyst decomposes to form typically phosphoric acid.

2. The phosphoric acid reacts with the carbonific compound to give a phosphate ester.

3. The phosphate ester decomposes to phosphoric acid, water, and a carbonaceous char.

4. The resin binder softens or melts, covering the carbonaceous char and trapping the released gases.

5. The spumific compounds decompose, releasing nonflammable gases that blow the carbonaceous char into the insulative foam protecting the substrate.

There is recent interest in the protection of flammable plastic substrates by the use of intumescent coatings. Slysh (1974) reported on the fire retardance of several plastics, including nylon and diallyl phthalate, using a conventional intumescent coating consisting of a vinyltoluence-butadiene resin binder, an ammonium polyphosphate intumescent catalyst, plus dipentaerythritol carbonific compound, together with melamine and chlorinated paraffin spumific compounds. When a 5-mil-thick dry film of this intumescent coating was applied to a glass reinforced nylon substrate, the flame spread, as measured on the 2-foot Monsanto tunnel, decreased from a rating of 140 for the uncoated control to 36. The protection afforded by the intumescent foam also can be measured by the insulative values of the control and coated nylon substrate. The insulative value is measured by the temperature increase on the back side of the panel. The lower the increase, the greater the insulative value. An uncoated glass-filled nylon sample had a back side temperature more than 400°C higher than a similar panel coating with a 5-mil-thick dry film. This effect illustrates the thermal protection afforded to the substrate by the intumescent foam.

2.7.4 Nonconventional Intumescent Coatings

By definition, nonconventional intumescent coatings are those in which the elements of intumescence are built into the binder itself.

A novel, clear and efficient intumescent epoxy coating has been prepared by the reaction of triphenyl phospite with an epoxy resin prepared from epichlorohydrin and bisphenol A. The coating, consisting of 100 percent solids, was prepared by adding the amine catalyst to a premised epoxy-(triphenyl phosphite) resin just before it was applied. It is proposed that triphenyl phosphite transesterifies a hydroxyl group in the epoxy resin. The epoxy becomes crosslinked not only by the polyamine curing agent, but also by the phosphite.

The phosphite is chemically bound into the resin; immersion of the coating in distilled water at 50°C for 24 hours did not affect its fire retardant properties. A flame spread of 37, as measured by a modified ASTM E84 test, was obtained with this resin when applied as a 7.6-mil-thick dry film.

2.8 Conclusions and Recommendations

2.8.1 Introduction

This section summarizes the conclusions and recommendations reached by the committee after having reviewed the state of the art of polymeric materials with respect to fire safety. These conclusions and recommendations are not intended to reflect any order of importance or priority. Moreover, the committee did not feel that it should provide quantitative statements of effort associated with each recommendation.

These conclusions and recommendations have been subdivided into three subsections. The first subsection presents summary conclusions upon which no specific recommendations are based. The second lists general conclusions and recommendations based on them. Finally, specific conclusions and recommendations concerned with particular materials areas are presented in the third subsection.

2.8.2 Summary Conclusions

Given sufficient oxygen and heat input, all organic polymers will burn.

Absolute fire safety of polymeric materials does not exist. There are always trade-offs in safety, utility, and costs.

Billions of pounds of synthetic and natural polymers are used annually in the United States without presenting unmanageable fire safety problems. However, some uses of polymeric materials have seriously augmented the fire hazard. With increased volume and diversity of uses, such hazards could increase further.

Many synthetic organic polymers burn in a manner differing from that of the more familiar natural polymers such as wood, paper, cotton, or wool. Some synthetics burn much faster, some give off much more smoke, some evolve potentially noxious and toxic gases, and some melt and drip. Others burn less readily than the natural polymers.

As the diversity and amount of polymeric materials used in confined areas (dwellings, vehicles, etc.) increases, the problems presented by the generation of smoke and toxic gases in a fire also increase. Objective information fully defining the extent and nature of this hazard is not available.

The fire safety of many polymers has been improved by the incorporation of fillers and/or compounds containing halogens, phosphorus, and/or antimony. In general, this is effective in resisting small ignition sources or low thermal fluxes in large-scale fires. Most of these systems burn readily and may lead to increased smoke generation or increased production of toxic and/or corrosive combustion products.

2.8.3 General Conclusions and Recommendations

Conclusions: Assessment of the fire safety characteristics of polymeric materials is hampered by a lack of adequate test methods and performance criteria. Many current developmental efforts simply seek to improve the fire performance of well-known materials so that they comply with existing and often inadequate standards (also see Volume 2 of this series of reports). Recommendations: Define and implement a broadly based program to develop meaningful test methods and standards that will more accurately define the flammability characteristics of polymeric materials in the broad spectrum of fire situations. Develop tests and protocols for ascertaining the safety hazards of combustion products of polymeric materials and materials combinations.

Conclusion: The fire safety characteristics of polymeric materials must be viewed in a systems context. Fire safe design, fire detection, fire fighting, the end use, environmental effects and general fire consciousness are at least equally important aspects of the total picture. Methods for fire safety analysis are ill-defined and/or seldom used (also see Volume 4 of this series of reports). Recommendation: Develop fire safety systems analysis methods to guide materials selection so that the overall fire safety assessment can be based on the material, design, environment, detection, and fire fighting factors.

Conclusion: The planning, coordination, and dissemination of research efforts by government agencies need improvement, and there is no central data bank on fire safety with unrestricted access to the fire research community. Recommendation: Create an overall program that will coordinate the government-supported work on fire safety of polymeric materials and disseminate its results.

Conclusion: Designers, architects, and engineers generally lack training regarding the properties of polymeric materials they use and specify. They are usually uninformed about the fire safety problems associated with these materials. Recommendations: Require greatly increased emphasis on fire safety of polymeric materials in curricula for designers, architects, and engineers; continue efforts to eliminate misrepresentation of fire safety in promotional materials and advertisements.

Conclusion: The general public lacks understanding of the fire safety aspects of polymeric materials. Because of the mode of burning, some synthetics may differ from the more familiar natural polymers. Those synthetics often present unexpected fire situations with which the average person cannot cope. Recommendation: Develop a sound educational program for all age levels to better acquaint the public with the fire safety aspects of polymeric materials, their products, and their uses.

Conclusion: Apprehension exists that the fire retardation of polymers in nonfire and/or fire situations may have an adverse impact on the environment. Recommendation: In all programs, maintain an awareness of the potential threat to the environment of additives, monomers, degradation products, and other materials used in the synthetics and modification of polymers for improved fire safety characteristics.

Conclusion: There is a serious lack of knowledge concerning the relationships of chemical and physical composition of polymeric products to the emission of smoke and toxic gas formation during combustion. Recommendation: Initiate programs to determine the relationship of chemical and physical components of polymeric products to the evolution of smoke and toxic gas formation.

Conclusion: There is a lack of basic knowledge concerning the relationship between the chemical and physical properties of polymers and fire dynamics parameters (flame spread, ease of ignition, etc.) as well as the way these relationships are affected by aging (see Volume 4 of this series of reports). Recommendation: Initiate programs to further increase basic knowledge of the relationships between the chemical and physical properties of polymers and fire dynamics parameters and the way these relationships are affected by aging.

Conclusion: Polymeric materials with improved fire safety characteristics that are available today have one or more deficiencies such as high cost, fabrication difficulties, and toxic and corrosive combustion products. By using polymers with appropriate molecular structures, it has been shown that one can produce polymers with fire safety characteristics that meet most end-use requirements. To gain large-scale acceptance, these materials must become available at reasonable cost. Recommendation: Support approaches to improving the fire safety of the high-volume, low-cost polymers (see Volume 1, Table 7). Establish incentives to accelerate the introduction and commercialization of new materials with improved fire safety characteristics. Increase the development effort on char-forming systems with particular emphasis on lowering the fabrication cost.

2.8.4 Specific Conclusions and Recommendations

Conclusion: Because of its low cost, comfort, and versatility, elastomeric polyurethane cushioning is widely used. However, such cushioning can burn readily, even when fire-retarded. New materials and/or new approaches using current materials are urgently needed. Recommendation: Institute and support research and development programs aimed at improving the fire safety characteristics of cushioning systems. Such programs could consider cost effectiveness *inter alia.*

Conclusion: Mass-market textile fibers (see Volume 1, Table 3) do not offer protection against direct exposure to flames. The fire safety aspects of fiber blends cannot be predicted from a knowledge of the behavior of individual components (e.g., effective and economical fire-retardant treatments for popular blends of cotton and polyester fibers are still the subject of considerable research). Carpet systems vary greatly in their response to fire. Systems that are safe for uses in critical areas such as aircraft and nursing homes need to be defined and/or developed. Recommendation: Institute and support a program to define the overall fire safety problems (flammability, smoke and toxic gases, etc.) of fiber-based materials and to develop improved materials as appropriate.

Conclusion: Concern exists about potential fire hazards associated with the rapidly increasing use of polymeric structural and insulating foams, particularly in the areas of kitchen cabinets, furniture applications, and decorative paneling.

Recommendations: Define and implement a broadly based program to develop meaningful test methods and standards that will more accurately establish the flammability characteristics of these polymeric systems in the broad spectrum of fire situations. Develop fire safety systems analysis methods to guide the materials selection process so that the overall fire safety assessment of these foams considers composition, end use, material design, environment, detection, and fire fighting factors.

Conclusion: The use of intumescent coatings can be a cost-effective way to enhance the fire safety of some polymeric products. Recommendation: Expand the materials and applications studies on intumescent coatings with emphasis on lowering the cost and improving coating performance and aesthetics.

CHAPTER 3

TEST METHODS, SPECIFICATIONS & STANDARDS

3.1 Introduction

Flammability of polymeric materials and fire hazard posed by a given product depends greatly on many factors in addition to composition. Geometry, orientation, ventilation, proximity of other materials, etc., can have overriding effects on the response of a given material in a fire.

If laboratory measurements of relative flammability of a given material could be interpreted and analyzed to enable prediction of the behavior of the material under real fire conditions, progress would be accelerated. Unfortunately this has not been the case. Understanding of fire phenomena has not been sufficient to provide the basis for valid simulations in laboratory tests.

The "state-of-the-art" is reviewed here in a general way. Technical problems of fire testing in polymeric materials, and some of the reasons for lack of correlation between results of laboratory tests behavior of materials under actual fire conditions are discussed.

3.2 Overview of Test Methods, Standards, Specifications and Codes

In the United States the role of materials in fire safety is governed by a complex hierarchy of test methods, standards, specifications, codes and related regulations.

The objective of all fire safety activities is to reduce loss from unwanted fires. For the present purpose attention will be focused on two major components: human death and injury and the destruction of material property.

Fire safety activities fall into five general categories:

1. The unorganized efforts of individuals, based on limited knowledge of the problem and concern with the safety of their own property and the personal safety of their family and associates.

2. The voluntary efforts of associates concerned with the development of fire test methods and recommended safety practices and the dissemination of authoritative fire safety information.

3. The economically motivated efforts of industrial organizations, including the insurance industry, concerned primarily with minimizing property loss.

4. The statutory efforts of government regulatory agencies to reduce the frequency and severity of fires through mandatory codes and standards.

5. The organized efforts of the Fire Services directed primarily toward the prevention, control, and extinguishment of fires.

Test methods, standards, specifications, and codes have their origin and find their applications principally in activities (2) and (4) above.

TEST METHODS are methods for measuring a property or behavioral characteristic of a material, product, or assembly as an aid to predicting its performance in application. If it is recognized as a generally useful procedure, a method may be submitted to a voluntary standards writing and promulgating body such as the American Society for Testing and Materials (ASTM) or the National Fire Protection Association (NFPA) or to a technical society or trade association for possible adoption. After further consideration it may be accepted and documented as a standard test method.

The term "Standard" is applied to a variety of rules and procedures designed to provide an orderly approach to problems of fire safety and may include definitions and terminology, methods of measurement, performance requirements, and rules for the protection of persons and property. The single most important characteristic of a standard is that it is legitimatized by some form of due process.

A standard may specify a level of performance suitable to the proposed application, a test method to measure performance, sampling plan, record keeping requirements and whatever else may be necessary to demonstrate compliance with the standard.

SPECIFICATIONS find their principal use in commercial practice where they are used to define the properties of a material, product or structure being procured. They establish the level of performance which the items must meet and test method by which the performance is to be measured.

CODES are a compilation of standards and recommended practices designed to assure a desired level of safety in a building or building sub-system.

The United States does not have a nation-wide building code which can be used to govern fire safety among other aspects of building construction. Instead, these requirements are covered by some 20,000 local codes which possess local differences that pose a serious problem to architects, designers, material suppliers and builders.

3.2.1 Present Status of Fire Test Methods Applicable to Polymeric Materials

The field of fire hazard testing of materials is in the state of ferment at the present time.

Up to about 10 years ago the emphasis of fire safety regulations, fire test methods, and fire research was on the protection of property.

With the recognition that the human cost of fires, in terms of deaths and injuries, might equal or exceed the direct property loss, and with increased emphasis on the safety of the public, the main thrust of fire safety development has shifted from the protection of structures and property to the protection of building interiors and their occupants. This has focused attention on the fire safety properties of a new class of products.

Many of these new products are made of polymeric materials. Fire performance tests designed to evaluate the structural fire performance of traditional materials were inappropriate for these new materials and application. Many new small scale test methods were originated as ad hoc tests in development laboratories. Few of these have achieved recognition as standard test methods by virtue of their endorsement by ASTM, NFPA, Underwriters Laboratories and other recognized organizations.

The action of the Federal Trade Commission against the cellular plastics industry triggered a massive reexamination of the relevance of existing fire test methods to the prediction of the performance of materials, products, and systems in fires.

The FTC complaint alleged

> "The aforesaid ASTM test standards are neither reliable nor accurate tests for determining, evaluating, predicting or describing the burning characteristics of plastics products under actual fire conditions."

The crux of the complaint is found in the last three words — What are "actual fire conditions"? To consider this question ASTM established Committee E-39 on Fire Hazard Standards. (Recently ASTM Committees E-5 and E-39 merged as ASTM Committee E-5).

A basic shortcoming of most small scale tests can be illustrated by reference to a typical test method such as ASTM D-1692. Here a small sample of the test material in a horizontal orientation is ignited at one end and the progress of burning is observed. In the combustion of a solid fuel, energy feedback from the high temperature gaseous combustion zone pyrolyzes the fuel surface to provide a continuing supply of gaseous fuel to the flame. The rate of burning is directly related to the magnitude of this energy feedback. In the D-1692 test configuration, most of the energy of combustion is dissipated in the rising convective plume and through radiation to the cool surroundings. In a real fire, on the other hand, energy exchange between adjacent fuel surfaces and radiation from the heated surroundings greatly increases the energy feedback and the intensity of combustion. Experiments at Factory Mutual Research Corporation have shown that the burning rate of a wood crib in a well ventilated compartment may be nearly twice as great as that of a similar crib burning in the open because of increased energy feedback. Obviously, any test which pretends to simulate "actual fire conditions" must simulate energy environment of a real fire.

3.2.2 Development of Tests and Standards

In the past, test methods development and the formulation of codes and standards has proceeded in a largely unstructured and sometimes haphazard fashion. In more recent years, the process has begun to become more formalized, and new tests and standards are expected to demonstrate a rational relationship to the hazards they are designed to simulate and control.

The rational development of an effective fire safety standard is a four-step process:

1. Identification of Hazard
2. Quantification of Hazard
3. Development of Test Method
4. Development of Performance Standard

3.2.2.1 Identification of Hazard

The order of magnitude of the fire loss in this country is well known and this alone is sufficient to justify a major effort to reduce the loss.

A major source of fire incident data is the FIDCO (Fire Incident Data Organization) file maintained by the NFPA. The computerized data base contains information on approximately 30,000 fire-related incidents in the period 1971–1975. This is the largest source of U.S. data on major fires and is invaluable in the identification of hazard area and the establishment of priorities.

In an effort to establish a more statistically valid measure of fire experience, the National Bureau of Standards (NBS) and the Consumer Product Safety Commission (CPSC) jointly sponsored a National Household Fire Survey conducted by the Bureau of the Census. A statistically selected sample of 33,000 households were interviewed and a total of 2,463 fire incidents were reported to have occurred during the preceding year.

This effort is being continued by the recently established National Fire Data Center of the National Fire Prevention and Control Administration (NFPCA).

Ultimately the National Fire Data System, which will include data from the fire services surveys, and from other public and private sources, should provide a greatly improved source of fire information. However, such data collections can only provide retrospective information on past events. They cannot predict the future as new materials, new applications, and new construction techniques are introduced.

Another area where statistical surveys are of limited value is the infrequent catastrophic fire such as results from an aircraft crash or in an oil refinery. Here the sample is too small (fortunately) for valid statistical analysis. However, the scenario approach can provide a useful tool for the identification of hazard. (See 3.5 and Chapter 5).

3.2.2.2 Quantification of Hazard

To guide remedial action it may be necessary to know the precise materials or products which played significant roles in the fire, the nature of the ignition source, the growth pattern of the fire, the behavior of humans who may have been involved, the mechanism of injury or loss, and many other details of the incident.

To provide such needed information NBS established the Flammable Fabrics Accident Case and Testing System (FFACTS). This system provided indepth investigations of more than 3,500 fire incidents involving flammable fabrics. Samples of the materials involved were obtained wherever possible for identification, laboratory testing, and sometimes accident simulation. More than 100 attributes of a given incident could be coded for automated retrieval from the data bank (not all these data would be available from a given incident). Through analysis of this data file it was possible to identify such factors as the type of product most frequently involved, the sources of ignition, the types of fabrics, the age groups most frequently injured, and the cause of injury (burns or smoke and gas). It was then possible to develop test methods and standards directed to specific causes of loss.

3.2.2.3 Development of Test Method

With the identification and quantification of the hazard to be addressed, the development of a suitable test method to measure hazard potential can usually proceed in a straightforward manner. Malhotra has listed the following attributes of a well-designed fire test method.

"(a) Environmental conditions: The test should reproduce the heating regime source, thermal feedback, oxygen supply, movement and dispersal of combustion products as is likely to be experienced in practice.

(b) Range of applicability: The enviornmental conditions should be capable of variation to increase the applicability of the test.

(c) Material representation: The modeling of the material should be such as to exclude effects of size, the presence of joints and junctions.

(d) Flexibility: The test should be capable of reproducing different orientations in which the product can be used.

(e) Reproducibility, repeatability and discrimination, should be of an acceptable level depending upon the nature of the test. A variance of 5 percent is satisfactory in most cases, in some even 10–15 percent is acceptable.

(f) Ease of operation: Small tests should be capable of single-handed operation in no more than two hours; even complex tests should not take more than one day.

(g) Meaningful expression of results: The results should be expressed objectively in units which make comparison easy. Descriptive phraseology should be avoided."

A strict interpretation of item (c) is questionable since size must inevitably have an effect in real fires and the test method should give an indication of this effect. And, if joints and junctions are characteristic of the mode of application of the product, the behavior of these under fire conditions is a legitimate object of the test.

The reproducibility requirements implied in item (c) also appear to be optimistic and unnecessarily restrictive. Such precision can be achieved under carefully controlled laboratory conditions. However, experience with test methods which simulate more closely the conditions which might be expected in a real fire indicates that much larger variance is to be expected.

3.2.2.4 Development of Performance Standard

A final step in the technological portion of the process of reducing fire loss is the development of a standard or recommended practice which, when properly implemented, will reduce the probability of occurrence of the particular type of loss addressed. A good standard will consist of three parts: a test method by which the hazard potential of the material, product or system is to be measured; criteria which establish appropriate levels of performance for the given application (these may vary with the application for a given product); and sample plans, inspection procedures and other auxiliary requirements by means of which the producer, consumer, and regulator can be assured that the product does indeed meet the requirements of the standards.

The test method selected must be appropriate to the hazard to be controlled. Thus, in the case of a fire retarded cotton batting that performed well in an open flame ignition test smoldering combustion occurred when ignited by a cigarette. The test must predict this.

The establishment of suitable levels of performance is a critical step in the development of a standard. Too low a level may allow the continued existence of unacceptable hazards while too high a level will place an unreasonable burden on society in terms of increased cost and limitations on the choice of goods.

Formal cost benefit analysis has not been applied to the setting of performance levels because of the complexity of the problem and the difficulty of obtaining adequate data. Standards have usually been established on an intuitive basis, guided by accident experience, simulation experiments, and an estimate of the probable economic impact.

Having established suitable levels of performance and a means of measuring performance, it is necessary for the supplier to be able to determine that his product meets the requirements of the standard. The consumer requires assurance that the product he purchases will perform as advertised, and the regulator must have a procedure for policing the market place. Since fire hazard tests are almost invariably destructive tests, the concept of 100% inspection is obviously inapplicable. Some form of quality control or statistical sampling plan is required and has been addressed by researchers. It must be remembered that validation and legitimation are essential to any standard development.

3.2.3 Classification of Fire Test Methods

Strictly speaking, a fire test method is a method which can be used to predict the performance of a material, product, structure, or system under a fire exposure condition that can reasonably be anticipated in the intended application. Test methods differ widely in purpose, scale, degree of sophistication, and other attributes, making systematic classification difficult.

Hilado divides test methods into two groups: research tests and acceptance tests.

Further, test methods may measure different fire hazard characteristics such as:

ease of ignition
surface flame spread
heat release
smole evolution
toxic gas formation
fire endurance

Acceptance tests may be further classified according to the intended end-use of the product; this frequently requires compliance with specific regulations. Finally, tests are frequently classified according to size and designated by such descriptive and non-quantitative terms as large scale, full scale, subscale, small scale and laboratory scale.

Robertson classifies test methods as property tests and system tests, and makes a further distinction between nondestructive and destructive tests and between active and passive tests. Malhotra expands this list to include basic property tests, quality assurance tests, hazard assessment tests and ad hoc tests designed to deal with specific situations.

In the following sections test methods will be discussed under several of the headings suggested above without attempting a rigorous classification.

3.2.3.1 Research Tests

Research tests provide a better understanding of some particular aspect of fire behavior under well-defined conditions, rather than predicting product performance in a real fire.

They are characterized by careful control of the environment of the experiment, extensive data collection, and detailed analysis of the data.

Scaling and modeling experiments constitute a special class of research experiments. They are used because they may be less expensive, less hazardous, more reproducible and more amenable to precise measurement and analysis. The purpose of a modeling experiment is to develop rules for predicting the outcome of a prototype test from the results of a subscale test or experiment.

The complete scale modeling of a fire is impossible and all useful models represent cases of partial modeling where only selected aspects of the prototype are modeled exactly.

A simple example of the difficulties encountered can be found in the observation that small flames are usually laminar in character while real fires are almost always turbulent. Obviously, a small laboratory flame cannot be expected to model the combustion behavior of the same fuel in a turbulent fire environment.

Difficulties are also encountered in maintaining the ratio of radiative to convective heat transfer constant as the scale of the fire is changed.

In a more elaborate investigation of modeling techniques it was shown theoretically that many of the characteristics of a large-scale fire can be accurately modeled by a laboratory fire burning at an elevated pressure. Not all aspects of the fire are modeled successfully by the pressure modeling techniques, but it affords an excellent example of the application of partial modeling to the design of reduced scale experiments.

3.2.3.2 Property Tests

Property tests measure a property or performance characteristic of a material independent of the product or application in which it is to be used. They may be further subdivided into measures of intrinsic properties, independent of sample geometry and test environment (e.g., heat of combustion, thermal conductivity, heat capacity) and measurements of performance properties which reflect an interaction with the test environment under carefully standardized conditions (e.g., autoignition temperature, oxygen index, specific optical density).

Test methods of this type are useful in product development, quality control, establishment of specifications and regulatory activities.

Property tests are usually small in size and simple and inexpensive to carry out. They should be performed according to detailed and carefully documented procedures and the results should be capable of objective interpretation, free of any opportunity for subjective judgments.

3.2.3.3 System Tests

In contrast to property tests, system tests emphasize interactions – interactions between material properties and the configuration in which the material is used, interactions between the various components which make up a system, and interactions between the system and its environment. They are designed to simulate the significant features of an anticipated fire exposure under application conditions. Thus, they may be used to predict performance and to provide the basis for effective hazard control standards.

The world of real fires embraces an almost unlimited range of possible conditions. The following is a partial list of some of the variables which can have significant effects on the outcome of a system test or a real fire:

Ignition Source
Combustibles
Ventilation
Confinement

A single test represents only one point in a multidimensional space. It is seldom practical to conduct a sufficient number of tests to provide a reasonable sample of the fire conditions that can be anticipated. Therefore, it is extremely important to be able to generalize from the results of a particular test and predict the effects of limited variations in fire conditions. This is perhaps best accomplished through the scaling and modeling techniques referred to above, but it may also be accomplished empirically by testing at more than one level of the appropriate test variable.

System tests are most useful as test methods to be referenced in fire hazard standards and codes. When properly designed, they combine a practical level of operability with a demonstrable relationship to performance in real fires.

3.2.3.4 Prototype Tests

Prototype tests expose the fully developed product, system, or structure under test to a fire environment that may reasonably be anticipated under conditions of actual use. By definition, they are full-scale tests.

Prototype tests may be considered to be a limited case of the system test where the conditions which determine performance are not merely simulated but are following in exact detail. Prototype tests are apt to be very expensive, thus restricting the number of tests that can be conducted.

Circumstances sometimes provide opportunities for prototype tests (ad hoc tests) which would otherwise be prohibitively expensive. Thus, buildings scheduled for demolition have been utilized in a number of notable tests. Similarly, a bus which had suffered severe mechanical damage afforded an opportunity to conduct burnout tests on the vehicle's interior.

A prototype test represents only a single point in the fire matrix and the ability to generalize from the results is severely limited. An important benefit from a prototype test is confirmation of the predictions drawn from property and system tests made during the development of the item under consideration. In situations where a very high level of fire safety is required, for example, in a manned satellite or an aircraft, prototype testing is the only means presently available to provide the required level of assurance.

3.2.3.5 Size and Scale of Tests

It is possible to recognize the difference between *size* and *scale*. Size refers to the physical dimensions of the test assembly while scale refers to its size relative to that of the prototype. Prototype tests are, by definition, full-scale tests. System tests may be full-scale tests or subscale tests. Subscale tests are obviously smaller than their prototype and make use of suitable modeling relationships to relate the results to the expected performance of the prototype.

Small or laboratory size tests may be loosely defined as those which can be conducted in a conventional laboratory. Their characteristic dimensions will usually not exceed one meter and range downward to micro-scale property measurements. Large-size tests may be defined as tests which require a specialized test structure or are conducted in the open.

Full-scale tests provide the most reliable measure of hazard potential, but their use is frequently impractical from the standpoint of cost and convenience. Subscale tests can provide a practical alternative if they are carefully designed to model correctly the essential features of the prototype test.

3.2.4 Fire Development

Most fire tests and standards treat fire as a quasi-steady phenomenon. Many test methods subject a sample of material of small fixed dimensions to an arbitrary energy pulse and record the material's response. Even in cases where the test method embodies a rate concept (ASTM E-162, NFPA 258, etc.), the results are usually used in building codes and standards in form of integrated, non-time dependent material properties (flame spread index, specific optical density, heat of combustion, etc.).

Codes and standards do give limited recognition to the time dependence of fire phenomena through the use of time ratings for structural components when subjected to a programmed temperature-time history (ASTM E119, etc.). The Federal Aviation Administration has recently called attention to the concept of time dependent phenomena in fire standards through proposals to relate the allowable rates of smoke and toxic gas production to a "time to escape."

Real fires are transient phenomena that follow a well defined pattern.

Various stages in the fire development process are of significance in the control of fires and the limitation of loss. These stages are measured by the occurrence of critical events on a time line. Actually, two time lines must be considered. The first is defined by the progress of the fire and includes the time of ignition, the times to reach critical levels of temperature, smoke concentration, and concentration of toxic gases, the time to flashover, the time of structural failure, the time to the start of decay, and the time of extinction. The second is determined by the human (and mechanical) response to the fire and includes the time of detection, the time of activation of a fire extinguishing system, the time of evacuation, and the time of arrival of the fire department.

In real fires the exact time of ignition is seldom known. The rate of the later stages of fire growth is largely independent of the characteristics of the ignition source.

The growth rate will depend on the geometry and combustion characteristics of the fuel elements, the dimensions of the fire compartment, ventilation, and other parameters relating to a specific fire. The initial growth of a fire is due to the spread of the fire over the surface of the fuel element first ignited, followed by spread over other fuel elements. The fuel consumption rate will be equal to the product of the fuel density, the burning area, and the linear regression rate of the burning surface.

In a room or compartment, the stage at which all exposed fuel surfaces become involved is termed "flashover." At this point conditions inside the compartment are clearly untenable and ad hoc efforts at fire fighting are unlikely to be effective. Polymeric materials can play a very significant role in the development of flashover conditions since the fire, at this stage, involves the interior furnishings and finishings of the room, many of which are made of polymers, rather than the structural elements of the building.

Much effort has gone into the study of fully developed (post-flashover) fires and the development of test methods and standards for their control. In contrast, the period between ignition and flashover has received much less attention. With increased recognition of the role of interior furnishings and finishings in the early stages of fire growth, this area should receive much greater emphasis in the future. Improved test methods to characterize their contribution to fire growth and effective standards to guide and control their application are urgently needed.

3.3 Description of Individual Test Methods

3.3.1 Introduction

The tests described or listed in this section are either widely used or significant in the evaluation of fire safety characteristics. In addition to tests developed in the United States, some tests developed in other countries are included where they offer some promise of filling a need.

For ease of reference, tests are identified by ASTM and ANSI numbers where possible. Other tests are more fully identified. References include documents that describe their use in fire studies of polymeric materials.

3.3.2 Tests for Ease of Ignition

Ease of ignition may be defined as the facility with which a material or its pyrolysis products can be ignited under given conditions of temperature, pressure, and oxygen concentration.

3.3.2.1 ASTM E-136 and D-1929

These tests employ a vertical furnace tube heated by electrical current passing through nichrome wire in an asbestos sleeve wound around the tube, and an inner refractory tube inside which the specimen is placed. Air is admitted at a controlled rate, and its temperature is measured by means of thermocouples.

A material is considered noncombustible by this test (ASTM E-136) if specimen temperatures do not increase more than 30°C (86°F) and there is no flaming after the first 30 seconds.

In the ASTM D-1929 test, a three-gram specimen is observed. Flash-ignition temperature is defined as the lowest initial temperature of air passing around the specimen at which a sufficient amount of combustible gas is evolved to be ignited by a small external pilot flame. Self-ignition temperature is defined as the lowest initial temperature of air passing around the specimen at which, in the absence of an ignition source, the self-heating properties of the specimen lead to ignition or sustained glow. This test is also known as the *Setchkin* ignition test.

3.3.2.2 ASTM D-229

The ASTM D-229 test for electrical insulation materials consists of two parts. In part 1, a horizontal specimen is ignited at one end by a vertical specimen exposed to the heat from a coil of nichrome resistance wire with continuous sparking provided to ignite evolved gases.

3.3.2.3 Method 2023 of Federal Test Method Standard No. 406

This method employs a vertical specimen surrounded by heater coils supplied with 55 amp. of electrical current.

3.3.2.4 Underwriters Laboratories Ignition Test (General Use)

This test employs a high-temperature glass flask surrounded by a molten-alloy bath heated by an electrical furnace. Specimens are dropped into the flask after temperature has reached equilibrium at a selected value, and are checked for ignition.

3.3.2.5 Underwriters Laboratories Tests (Electrical Use)

Four of these tests are used to evaluate ease of ignition of materials intended for electrical applications. The hot-wire ignition test employs specimens in three thicknesses, wrapped with five turns of No. 24 wire. The wire is brought to red heat by applying a 65 watt current for a maximum of five minutes. The high-current-arc ignition test subjects a specimen to 40 applications per minute, for a maximum of five minutes of an electric arc. The high-voltage arc ignition test subjects a specimen to a 5200 volt arc for a maximum of two minutes. The high-voltage arc-tracking test subjects a specimen to a 5200 volt arc which is repeatedly extinguished and re-established by changing the arc distance, until the conductive path is 50.8 mm (2 in.) in length or until two minutes of arching time has elapsed.

3.3.2.6 Method 4011 of Federal Test Method Standard No. 406

This test measures the resistance of a specimen to arcs while exposed to 56 watts of heat.

3.3.2.7 Department of Commerce, DOC-FF-4-72 (Mattress Test)

The upholstered furniture test is similar to this test.

3.3.3 Tests for Surface Flame Spread

Surface flame spread may be defined as the rate of travel flame front under given conditions of burning. This characteristic provides a measurement of fire hazard, in that surface flame spread can transmit fire to more flammable materials in the vicinity and thus enlarge the fire even though the transmitting material contributes little fuel to the fire.

3.3.3.1 Twenty-Five Foot Tunnel Test (ASTM E-84)

This is perhaps the most widely accepted test for surface flame spread. It requires a specimen 25 feet long and 20 inches wide, mounted face down so as to form the roof on a 25-foot long tunnel, 17½ inches wide and 12 inches high. Flame spread classification is determined on a scale on which an asbestos-cement board is zero and a select-grade red oak flooring is 100. Fuel contribution and smoke evolution are determined on a similar scale.

3.3.3.2 Radiant Panel Test (ASTM E-162)

This test employs a radiant heat source consisting of a 12 X 18 inch vertically mounted porous refractory panel maintained at 670 ± 4°C (1238 ± 7°F). A specimen measuring 18 X 6 inches is supported in front of it with the 18-inch dimension inclined 30 degrees from the vertical panel. A pilot burner ignites the top of the specimen, 4¾ inches away from the radiant panel, so that the flame front progresses downward along the underside exposed to the radiant panel. The temperature rise recorded by stack thermocouples, above their base level of 400°F, is used as a measure of heat release.

3.3.3.3 Eight-Foot Tunnel Test (ASTM E-286)

This test employs a specimen 8 feet long and 14 inches wide, mounted horizontally to form the roof of a tunnel to give a lengthwise slope of 6 degrees and a sideways tilt of 30 degrees.

3.3.3.4 The Union Carbide Corporation Four-Foot Tunnel

A specimen measuring 47½ inches long and 7½ inches wide is mounted horizontally to form the roof of a tunnel 6¾ inches deep. The heat supply rate is 325 Btu per minute.

3.3.3.5 The Monsanto Company Two-Foot Tunnel Test

A specimen measuring 24 X 4 inches is inclined 28 degrees from the horizontal to form the roof of the enclosed tunnel, and ignited by a Fisher 3-900 burner supplied with a constant three-ounce-per-square-inch natural gas flow. The test method provides flame spread rating, fuel contribution, afterflaming, afterglowing, intumescence, insulative value, and smoke contribution.

3.3.3.6 The Schlyter Test

This test employs two specimens, each 12 inches wide and 31 inches high, held in a vertical position with their faces 2 inches apart. The bottom of the assembly is subjected for three minutes to the flame from either a bunsen burner with wing top delivering 37 Btu per minute, or a No. 4 maker burner with a special T-head delivering 291 Btu per minute.

3.3.3.7 Pittsburgh-Corning 30-30 Tunnel Test

A specimen measuring 30 X 3-7/8 inches is inclined 30 degrees from a horizontal position to form the roof of the enclosed tunnel. The flame source used as a Fisher 3-900 burner supplied with 2600 cc per minute of natural gas, which gives a 6-inch flame providing about 90 Btu per minute.

3.3.3.8 ASTM D-635 Test

This is perhaps the most widely used small-scale test for plastic materials. The specimen measures 125 X 12.5 mm (5 X 0.5 in.) by the supplied thickness and is held with the 125 mm (5 in.) dimension horizontal and the 12.5 mm (0.5 in.) dimension inclined at a 45-degree angle. One end is contacted for 30 seconds with a 25 mm (1 in.) high blue flame from a 3/8-inch diameter barrel bunsen burner.

3.3.3.9 ASTM D-1692 Test

This test is perhaps the most widely used small-scale test for cellular plastics. The specimen is supported on a horizontal hardware cloth support. One end is contacted for 60 seconds with a high blue flame from a bunsen burner.

3.3.3.10 The Underwriters Laboratories 94 Standard Test

This test employs essentially the ASTM D-635 test as its horizontal burning test for 94HB classification, and the ASTM D-1629 test as its horizontal burning test for 94HBF, 94HEF-1, or 94HEF-2 classifications.

3.3.3.11 The ASTM D-3014 Test

A vertical specimen is contacted for ten seconds with a 960°C (1760°F) flame from a bunsen burner. The specimen is enclosed in a vertical steel chimney lined with reflective aluminum foil to increase the severity of exposure.

3.3.3.12 The ASTM D-569 Test

A specimen measuring 450 X 25 mm (18 X 1 in.) is held vertically, and the bottom contacted for 15 seconds with a 25 mm (1 in.) high flame from a bunsen burner.

3.3.3.13 The ASTM D-1433 Test

A specimen 228 X 76 mm (9 X 3 in.) is mounted at a 45 degree angle in a test chamber. The bottom end is exposed to a 13 mm (0.5 in.) flame from a No. 22 hypodermic needle jet.

3.3.3.14 The ASTM D-757 Test

One end of a specimen measuring 121 X 3.17 by 12.7 mm (5 X 1/8 X 1/2 in.) is placed in contact with a Globar element maintained at 950°C (1742°F) for 3 minutes.

3.3.3.15 Federal Test Method Standard No. 406

Two test methods similar to ASTM tests are described. Method 2021 is similar to ASTM D-635. Method 2022 is similar to ASTM D-568, except that ignition by a burner flame is replaced by ignition with a pyroxylin fuse or benzene drop.

3.3.3.16 Test For Paints (ASTM D-1360)

This test employs a cabinet measuring 13¼ X 9 X 18¼ inches and a specimen measuring 6.4 X 152 X 305 mm (¼ X 6 X 12 in.). The ignition source is 5 ml. of absolute ethyl alcohol in a cup under the specimen.

3.3.3.17 Test for Paints (ASTM D-1361)

A fire shield measuring 203 X 279 X 762 mm (8 X 11 X 30 in.), open at the top, and a vertical stick measuring 25.4 X 25.4 X 406 mm (1 X 1 X 16 in.) painted on all four sides are utilized. The stick is wrapped with a wick soaked with 4 ml. of ethyl alcohol and ignited.

3.3.3.18 Test for Fabrics (ASTM D-1230)

A chamber measuring 268 X 216 mm (14.5 X 8.5 in.) X 256 mm (14 in.) high and a specimen measuring 51 X 152 mm (2 X 6 in.) mounted at a 45-degree angle are required. The ignition source is a 16 mm (5/8 in.) flame from a 26 gage hypodermic needle, applied for one second.

3.3.3.19 Test for Floor Coverings (ASTM-2859)

A test chamber measuring 305 X 305 X 305 mm (12 X 12 X 12 in.) and a specimen measuring 230 X 230 mm (9 X 9 in.) are used in this test. The specimen is placed on the floor of the chamber. The ignition source is an Eli Lilly No. 1588 methenamine tablet. This test is essentially the same as that described in flammability standards DOC FF 1-70 and DOC FF 2-70, for carpets and rugs.

3.3.3.20 Federal Test Method Standard No. 501a

This test for nontextile resilient floor coverings describes two test methods. In *Method 6411,* a specimen measuring 31.5 X 7 inches is supplied with 9.6 cu. ft./hr. propane and 150 cu. ft./hr. air for 240 seconds in a hood consisting of a horizontal flue measuring 38 X 8 X 6 inches and a vertical flue measuring 18 X 8 X 6 inches. In *Method 6421,* a specimen measuring 18 X 6 inches, inclined at a 60° angle, is exposed to a 670°C radiant panel and ignited at the upper end.

3.3.3.21 Underwriters Laboratories Floor Furnace

This test consists of a test chamber measuring 10 X 22 in. X 10½ ft. A specimen measuring 24 X 96 inches is mounted horizontally on the floor, and one end is exposed to a gas burner flame delivering 500 Btu/min. and an air velocity of 100 ft./min. for 12 minutes.

3.3.3.22 Test for Roof Coverings (ASTM E-108)

Three test procedures are described here. In the *Intermittent Flame Test,* a specimen measuring 3 ft. 4 in. X 4 ft. 4 in., mounted at a specified slope, is exposed to flame at 760°C (1400°F) and a 12 mph air current, intermittently according to a specified sequence. In the *Flame Spread Test,* a specimen measuring 3 ft. 4 in. X 13 ft., mounted at a specified slope, is exposed to flame at 760°C (1400°F) for 10 minutes or 704°C (1300°F) for four minutes and a 12 mph air current. In the *Burning Brand Test,* a specimen measuring 3 ft. 4 in. X 4 ft. 4 in., mounted at a specified slope, is exposed to a 12 mph air current and a burning brand.

3.3.3.23 ASTM E-69

This is a *Fire Tube Test* for treated wood and employs a specimen measuring 9.5 X 19 X 1016 mm (3/8 X ¾ X 40 in.), mounted vertically in a three-inch diameter fire tube and exposed to a 279 mm (11 in.) high blue flame from a burner.

3.3.3.24 ASTM E-160

This is a *Crib Test* for treated wood and employs a specimen consisting of 24 pieces each measuring 13 X 13 X 76 mm (½ X ½ X 3 in.), exposed to a 254 mm (10 in.) high blue flame from a meker burner.

3.3.3.25 Underwriters Laboratories Standard UL-214

Two tests are described. In the *Small-Flame Test,* a specimen measuring 2¾ X 10 inches is mounted vertically and the lower end is exposed to a 1.5 inch long vertical flame from a burner for 12 seconds, in a test chamber measuring 12 X 12 X 30 inches. In the *Large Flame Test,* specimens in single sheets (measuring 5 X 84 inches) or folds (measuring 25 minimum X 84 inches, folded in 5-inch pleats) are mounted vertically, with the lower end exposed to an 11-inch long flame from a burner for two minutes, in a test chamber measuring 12 X 12 X 84 inches.

3.3.3.26 Federal Test Method Standard No. 191

Six methods for textiles are described in this standard. These are summarized as follows: *Method 5900* employs a specimen 7 X 10 inches, horizontal, exposed to vertical flame from 0.3 ml. anhydrous ethyl alcohol. *Method 5903* – A specimen 2¾ X 112 inches, vertical, lower end is exposed to 1½ inch long vertical flame from a burner for 12 seconds, the test chamber measures 12–14 X 12–14 X 30 inches. *Method 5904* – The specimen 2 X 5 inches, vertical, bottom edge is exposed to a vertical flame from ¾-inch diameter paraffin candle. *Method 5906* – The specimen 4½ X 12½ inches, horizontal, is exposed to 1½ inch long vertical flame from a burner, the test chamber measures 15 X 8 X 14 inches. *Method 5908* – The specimen measures 2 X 6 inches, inclined at a 45 degree angle, with the lower end exposed to 5/8-inch flame from a hypodermic needle for one second. *Method 5909* – A specimen 1 X 6 inches, is inclined at a 30 degree angle, with the lower end exposed to a flame from a safety-book match for five seconds.

3.3.3.27 Fabric Flammability Test

This test is described by the DOC FF 3-71 Flammability Standard For Children's

Sleepwear and employs a vertical specimen measuring 3.5 inches X 10 inches, mounted in a test chamber measuring 12 X 12 X 31 inches, and exposed to a 1.5 inch long vertical flame from a burner for 3.0 ± 0.2 second.

3.3.3.28 Motor Vehicle Safety Standard No. 302 Test

This test employs a specimen measuring 4 X 14 inches, mounted horizontally in a test chamber measuring 15 X 8 X 14 inches. One end is exposed to a 1.5 inch high vertical flame from a burner.

3.3.4 Tests for Rate of Heat Release

3.3.4.1 The Conventional Calorimeter

The conventional calorimeter (e.g., the Parr Bomb) measures the thermochemical quantity of heat released when a known small quantity of fuel is completely combusted in oxygen and the products of combustion cooled down to the initial fuel-air mixture temperature.

3.3.4.2 The Factory Mutual Construction Material Calorimeter

This calorimeter has been in use for a number of years for evaluating floor, roof, and wall assemblies and interior finishing materials. The specimen is exposed in the roof of a furnace. The exposure imposed on the sample is obtained from burning a heptane-air mixture. Interior finish materials are exposed for ten minutes, while construction assemblies are exposed for 30 minutes. The flue gas temperature is recorded throughout the test run. In a subsequent evaluation run, an inert (or reference) sample is substituted for the test sample and given the identical exposure. The flue gas time-temperature curve is reproduced by burning propane to make up the difference between the test and reference samples. The rate of heat release of the test sample is thus obtained from the rate of consumption of propane during the substitution run. It is the only calorimeter which imposes on a specimen a time-varying heat flux.

3.3.4.3 The National Bureau of Standards Heat Release Rate Calorimeter

This calorimeter employs a combustion chamber with inside dimensions of 0.33 X 0.33 X 0.36 m, lined on three sides with gas fired radiant panels whose temperature is varied between 900 and 1300°K to produce the desired irradiance of the specimen. The specimen is oriented vertically at the center of the combustion chamber, with only its front surface exposed to the radiant panels, its edges shielded by the insulated holder, and its back surface cooled by a water cooled brass block which is close to it but not in direct contact. The temperature rise in the water stream permits calculation of heat released in that manner. This configuration represents a section of a burning wall where the back surface of the wall is exposed to a relatively cool surface.

This instrument is designed to burn pyrolysis products completely by means of a propane diffusion flame in a secondary burner.

The combustion chamber is open at the top, allowing the hot combustion gases to pass freely into the control chamber above it. The control chamber is 0.45 m high and 0.33 X 0.33 m in cross section, contains an auxiliary burner, the gas flow to which is automatically controlled so that the average temperature of the gases passing up into the mixing chamber above remains constant. The propane flow is a direct measure of the rate of heat release.

The mixing chamber is above the auxiliary burner and contains a system of sheet metal baffles for mixing and combustion gases and four thermocouples to measure an average temperature.

3.3.4.4. The Stanford Research Institute Heat Release Rate Calorimeter

This calorimeter is a replica of the NBS calorimeter in basic design and scaled to take a sample 16 times larger. Both the radiant heating panels and the secondary burner are fired with a premixed natural gas flame. Its features include a wide range of radiant flux and the capacity to accept large size samples including model wall and other construction assemblies. By installing a smoke meter in the stack and using a by-pass around the secondary burner, smoke measurements may be made.

3.3.4.5 The Ohio State University Heat Release Rate Calorimeter

This calorimeter employs a chamber with a pyrimidal top section containing an electrically heated ceramic radiant panel. The instrument accepts a sample in the vertical and horizontal position.

The instrument can only attain moderate exposure levels and thus requires a pilot flame. The pilot flame is applied continuously. It was constructed to have a low heat capacity and low heat losses so that the rate of heat release from a sample could be calculated from the rate of temperature rise in the well-mixed products of combustion.

This calorimeter differs from the others in that it was designed to permit the simultaneous measurement of flame spread rate, heat release rate, smoke release rate and toxic gas production rate as a material is exposed to a known heat flux. It was intended to be used to evaluate the "life hazards" of materials during the initiation and flame spread period of the fires, and was therefore designed with low heat flux exposure capability.

3.3.4.6 Forest Products Laboratory Heat Release Rate Calorimeter

This calorimeter operates on the substitution principle, similar to the FM Construction Materials Calorimeter in 3.4.4.2, and requires two runs to evaluate each sample. It employs a gas-fired radiant panel which is only capable of low to moderate heat fluxes. The instrument employs a water-cooled sample holder which can accept relatively large size samples, including models of construction assemblies.

3.3.4.7 The ASTM E-84 25-Foot Tunnel Test and the ASTM E-162 Radiant Panel Test

These are essentially tests for surface flame spread, but they provide some measure of heat release based on the temperature rise in the gases produced. However, these tests lack assessment of rate, and only provide information relative to red oak (ASTM E-84) or hardboard (ASTM E-162) under specific test conditions.

3.3.5 Tests for Oxygen Requirements

A variety of research tests have been used in which the burning of a test specimen is observed in atmosphere containing varying amounts of oxygen. None of these tests achieve the status of a standard test method. Tests of this type may be used to assess the hazard potential of materials in hazardous oxygen enriched-atmospheres.

3.3.5.1 Fire Hazards In Oxygen-Enriched Atmospheres; NFPA No. 53-M

3.3.5.2 Flammability, Odor and Offgassing Requirements and Test Procedures for Materials in Environments that Support Combustion: NASA NHB8060.1A, February 1974.

3.3.6 Tests for Ease of Extinguishment

While test procedures exist to evaluate effectiveness of portable extinguishers on certain standardized fires (wood pellets, jeptome pools), the inverse procedure of using a standardized extinguisher to test a variety of materials has never been developed.

3.3.7 Tests for Smoke Evolution

Smoke density may be defined as the degree of light or sight obscuration produced by the smoke from burning materials under given conditions of combustion.

Some measures of smoke density are: degree of light absorption, specific optical density, and smoke development factor.

Tests for smoke evolution generally involve measurement of the fraction of light absorbed or obstructed by smoke evolved from a decomposing or burning material.

3.3.7.1 The National Bureau of Standards Smoke Density Test Chamber

This chamber is a completely closed cabinet, measuring 3 feet wide, 3 feet high, and 2 feet deep, in which a specimen 3 inches square is supported vertically in a frame such that 2-9/16 inches square is exposed to heat under either flaming or nonflaming (smoldering) conditions. The heat source is an electric furnace. A vertical photometer path for measuring light absorption is employed to minimize measurement differences due to smoke stratification which could occur with a horizontal photometer path at a fixed height.

Smoke measurements are expressed in terms of specific optical density (DS), a dimensionless value that presents smoke density independent of chamber volume, specimen size, or photometer path length.

The NBS test provides additional information, including maximum smoke accumulation, maximum smoke accumulation rate, time to reach maximum smoke density, and time to reach a critical smoke density.

The NBS smoke density chamber is the best equipment available for smoke evaluation, and is a candidate for adoption as a national standard.

3.3.7.2 The ASTM D-2843 Test

This test employs a cabinet measuring 30 inches high, 12 inches wide, and 12 inches deep, completely enclosed except for one-inch high openings around the bottom. The heat source is a propane-air flame from a Benz-O-Matic TX-1 pencil-tip burner, applied at a 45° angle to the bottom of a horizontal specimen. A horizontal photometer path, 20 inches above the bottom of the chamber, is used for measuring light absorption.

3.3.7.3 The ASTM E-84 Test

This is primarily a test for surface flame spread, but it provides a measure of smoke evolution by light absorption readings during the test.

3.3.7.4 The ASTM E-162 Test

This is a test primarily for surface flame spread, but it provides a measure of smoke evolution by collecting a smoke deposit by vacuum for subsequent weighing.

3.3.7.5 The Arapahoe Smoke Test

With this test, smoke evolution is measured by exposing a small specimen of material to a burner flame and collecting the smoke deposit for subsequent weighting. Small sample size and low heat flux limit the capability of this test.

3.3.8 Tests for Toxic Gas Emission

Tests for fire gases generally fall into two types: 1) those concerned with identification and analysis of the chemical compounds in the gaseous combustion products, and 2) those concerned with studying the effects of these gases on laboratory animals.

The chemical methods of identification and analysis that may be useful for studying fire gases are:

1. Infrared analysis
2. gas chromatographic analysis
3. mass spectrometric analysis
4. colorimetric indicators

Toxocological studies on laboratory animals generally involve exposure of small animals, usually rats or mice, to the gaseous products of decomposition and combustion under carefully controlled conditions, followed by autopsies and examinations to determine cause of death, if fatalities occur, and extent of injuries.

A test procedure for toxicological studies is used in Germany at the hygiene institutes of universities at Hamburg, Munster, and Aix-la-Chapelle, and at the toxicology institutes of Bayer (Wuppertal), BASF (Ludwigshafen), and Hoechst (Frankfurt). The apparatus consists of a horizontal fused silica tube with a length of at least 1000 mm, and outside diameter of 40 mm, and a wall thickness of 2 mm. An annualar electric oven tightly enclosing the tube is moved along the tube. The air maintaining the pyrolysis process is introduced into the tube through a flowmeter. The oven moves against the air stream, and the speed of the annular oven determines the burning rate of the material, and, together with the length of the test sample, determines the time of test duration.

A test procedure developed at the University of Tokyo exposes a specimen measuring 18 X 18 cm to radiant heat from a 1.5 kilowatt electric heater and to a propane gas flame from a burner fed with 0.35 liter of propane and 3 liters of air per minute. The combustion gases enter a dilution chamber for adjustments to suitable temperature and concentration, and are then led into an exposure chamber containing eight mice in movement detecting devices. Four of these devices are of the revolving type (with a freely revolving treadwheel) and four are of the strain type with a strain gage).

A smoke toxicity apparatus developed at Harvard Medical School consists of three major units and connecting pipes:

1. A combustion chamber, consisting of a stainless steel box centered by glass fiber insulation in a sheet metal outer box, and containing a specimen holder positioned above the base and equidistant from all sides. Energy input is provided by two 1750 W electrical heating elements with continuously variable output.

2. An animal mortality chamber partitioned to contain 16 Sprague-Dawley rates.

3. An animal time-of-useful-function chamber, containing ten wire mesh exercise wheels, each to contain one rat, driven at a speed to 10 rpm by a variable speed gear motor.

The entire apparatus has a volume of 300 liters. The test procedure measures animal mortality and time to death for the 16 animals in the mortality chamber, and time of useful function and time of unconsciousness for the 10 animals in the time-of-useful-function chamber. Exposure time is 15 minutes.

A screening test procedure developed at the University of San Francisco involves pyrolysis of materials at a heating rate of 40°C/min. in a horizontal tube furnace to a temperature of 500°C or 800°C. The animal exposure chamber is a 4.2 liter patented NASA design, and contains four freely moving Swiss albino male mice which are observed continuously for signs of incapacitation and death. Materials are ranked on the basis of time to incapacitation, time to death, and apparent lethal concentration of pyrolysis gases for 50 percent of the animals (ALCsub50). The test procedure can employ either pyrolysis in inert or oxidative environments. Exposure time is 30 minutes.

A 142-liter hyperbaric chamber is used at NASA Johnson Space Center and University of Tennessee to expose 10 rats to the pyrolysates from materials pyrolyzed at temperatures up to 800°C. The animals are introduced through an animal entry port after the pyrolysis products have been drawn into the chamber by the initial vacuum. Exposure time is 30 minutes.

Fabrics are burned with a flame and the combustion products passed into a 642-liter chamber at the University of Tennessee. The sample is exposed to the burner for three minutes, and the 10 rats in the chamber are exposed for a two-hour period.

A dynamic slow pyrolysis procedure at the University of Tennessee involves pyrolysis of the material in an air stream which carries the pyrolysis products into a 63-liter chamber containing four rats. Air flow is 1.0 liter/minute before the pyrolysis tube and 0.5 liter/minute after the pyrolysis tube.

3.3.9 Tests For Fire Endurance

Fire endurance may be defined as the resistance offered by the material to the passage of fire, normal to the exposed surface over which the flame spread is measured. This characteristic provides a measure of fire hazard in that a material which will contain a fire represents more protection than one which will give way when all other factors are the same. Some measures of fire endurance are penetration time and resistance rating.

3.3.9.1 ASTM E-119 Test Standard for Building Construction and Materials

This standard test provides for exposure of various structural components to a standard fire, the character of which is determined by a standard time-temperature curve.

The performance is defined as the period of resistance to standard exposure elapsing before the fire critical point in behavior is observed, and is expressed in time ratings, such as two-hour and four-hour.

For bearing walls and partitions, the area exposed to fire shall not be less than 100 square feet (9 sq. m), with neither dimension classification is desired, the wall or partition should not permit passage of flame or gases hot enough to ignite cotton waste, should sustain specified loads during and after test, and should not permit the temperature rise on the unexposed surface to exceed 139°C (250°F).

For nonbearing walls and partitions, the area exposed to fire shall not be less than 100 square feet (9 sq. m), with neither dimension less than 9 feet (2.7 m). For a period equal to that for which classification is desired, the wall or partition should not permit passage of flame or gases hot enough to ignite cotton waste, and should not permit the temperature rise of the unexposed surface to exceed 139°C (250°F).

For columns, the length of the column shall not be less than 9 feet (2.7 m). For a period equal to that for which classification is desired, the column should sustain a specified load.

For protected structural steel columns, the length of the column shall not be less than 8 feet (2.4 m). For a period equal to that for which classification is desired, the average temperature of the steel should not exceed 538°C (1000°F) at any of four levels, and the temperature at any of at least three points on any of the four levels should not exceed 549°C (1020°F).

For floors and roofs, the area exposed to fire shall not be less than 180 square feet (16 sq. m), with neither dimension less than 12 feet (3.7 m). For a period equal to that for which classification is desired, the construction should sustain a specified load, should not permit passage of flame or gases hot enough to ignite cotton waste, and should not permit the temperature rise of the unexposed surface to exceed 139°C. (250°F). For specimens employing steel structural members, the temperature of the steel shall not have exceeded 704°C (1300°F) at any location during the classification period, nor shall the average temperature recorded by four thermocouples at any section have exceeded 593°C (1100°F). For specimens employing concrete structural members, the average temperature of the tension steel at any section shall not have exceeded 427°C (800°F) for cold drawn pre-stressing steel or 593°C (1100°F) for reinforcing steel during the classification period.

For loaded beams, the length of beam exposed to fire shall not be less than 12 feet (3.7 m). A section of a representative floor or roof construction not more than 7 feet (2.1 m) wide may be included with the test specimen. For a period equal to that for which classification is desired, the specimen shall have sustained the specified load.

For protected structural steel beams and girders, the length of beam or girder exposure to fire shall not be less than 12 feet (3.7 m) and the exposed section of floor construction shall not be less than 5 feet (1.5 m) wide. For a period equal to that for which classification is desired, the average temperature of the steel at any one of four sections should not exceed 538°C (1000°F), and the temperature at any one of at least four points at any of the four sections should not exceed 649°C (1200°F).

For ceiling construction, the area exposed to fire shall not be less than 180 square feet (16 sq. m), with neither dimension less than 12 feet (3.7 m). For a period equal to that for which classification is desired, the ceiling should not permit the passage of flame, should not permit the temperature rise of combustible supports to exceed 139°C (250°F) at points of contact or at adjacent surfaces, and should not permit the average temperature of noncombustible supports to exceed 538°C (1000°F).

For protected combustible framing or facings, the area exposed to fire shall not be less than 100 square feet (9 sq. m) for wall and partition protection and not less than 180 square feet (16 sq. m) for floor protection, neither dimension being less than 9 feet (2.7 m) and 12 feet (3.7 m), respectively. For a period equal to that for which classification is desired, the protection should not permit ignition of the materials protected, and should not permit the temperature rise at points of contact with the protected structural members to exceed 139°C (250°F). The permissible temperature rise is 181°C (325°F) for members closely imbedded on three sides in noncombustible materials.

The standard time-temperature curve described in ASTM E-119 is also used to define the exposure fire for the ASTM E-152 Test for door assemblies and the ASTM E-163 Test for window assemblies.

Similar severity of exposure is involved in the Union Carbide *Corporation Tests for Thermal Insulation Materials for Pipes and Vessels.* In these tests, the flow of gasoline fuel is controlled so that the fire exposure approximates that of the ASTM E-119 standard time-temperature curve.

3.3.9.2 The Factory Mutual Heat Damage Test

A horizontal specimen measuring 16 X 16 inches of the roof or wall construction to be tested is employed. The specimen forms the roof of a small furnace, which is heated so that the temperature one inch below the bottom face of the specimen increases as follows:

Roof Insulations		Wall Insulations	
Temperature	Elapsed Time	Temperature	Elapsed Time
28°C (50°F)	at 10 min	236°C (425°F)	at 5 min.
28°C (50°F)	at 20 min	264°C (475°F)	at 10 min.
28°C (50°F)	at 30 min.	277°C (500°F)	at 15 min.
		277°C (500°F)	at 20 min

3.3.9.3 The Underwriters' Laboratories Test For Air Ducts

This test employs a horizontal specimen 18 inches square, subjected to an eight-pound static load with a bearing surface on the sample of 1 X 4 inches. The specimen forms the roof of the test furnace. Gas burners 20 inches below the underside of the specimen set at 760°C (1400°F) for a period of 30 minutes.

3.3.9.4 The SS-A-188b Test

A specimen measuring 36 X 36 inches is supported in a horizontal position with a 30 X 30 inch area exposed, 28¾ inches above a burner delivering 28,000 Btu during a 20-minute test of 60,000 Btu during a 40-minute test.

3.3.9.5 The Bureau of Mines Flame Penetration Test

This test employs a horizontally mounted specimen measuring 6 X 6 inches by one inch, and measures the time required for the one-inch thickness to be penetrated by a 1175°C (2150°F) propane-air flame from a vertical Benz-O-Matic pencil flame burner. An earlier version employed a vertically mounted specimen and a horizontal flame at 1930°F.

3.3.9.6 The NASA-Ames T-3 Thermal Test

Specimens measuring 12 X 12 X 2 inches are mounted in one of three test areas in the chamber. The flames from an oil burner supplied with approximately 1½ gallons per hour of JP-4 jet aviation fuel provide heat flux on these three areas of 5.5 to 11, 9 to 16, and 22 Btu per square foot per second.

3.3.9.7 The ICI Test For Pipe Insulation

This test employs a horizontal pipe 107 to 120 cm long, filled with an inert liquid and covered with the insulation material to be tested. The fire source is two liters of kerosene poured into a metal tray below the test pipe and ignited.

3.3.10 Full Scale Tests

Full scale fire tests are designed to reproduce actual fire scenarios under controlled and measured conditions. They can be divided into three types: (1) corner, (2) compartment, and (3) corridor tests.

3.3.10.1 The Eight-Foot Corner Test

This test was developed at the University of California (Berkeley) and employs enough specimen material to construct a corner consisting of two walls each 6-feet wide and 8-feet tall, topped with a 6-foot square ceiling. Various ignition sources have been used, such as a 2800-gram quantity of milk cartons.

This test configuration has been employed at Underwriters Laboratories, National Bureau of Standards, Union Carbide Corp., and other laboratories.

3.3.10.2 The Factory Mutual Corner Test

Enough sample material is employed to construct a corner consisting of two walls 24-feet, 9-inches in length. The igniting fire is normally produced by burning a 5-foot high pile of 4 X 4-foot wood pallets weighing about 750 pounds with an average moisture content of less than eight percent.

Compartment tests are important, but have not been standardized. Compartment test facilities in the United States include:

Battelle Laboratories, Columbus, Ohio
Factory Mutual Research Corp., Norwood, Massachusetts
Boeing Company, Seattle, Washington
McDonnell-Douglas Corp., Long Beach, California
National Bureau of Standards, Gaithersburg, Maryland
ITT Research Institute, Chicago, Illinois
Southwest Research Institute, San Antonio, Texas
Stanford Research Institute, Menlo Park, California
University of California, Berkeley, California
Underwriters Laboratories, Northbrook, Illinois
University of Washington, Seattle, Washington

Corridor test facilities with adjacent compartments, includes:

Factory Mutual Research Corp., Norwood, Massachusetts
ITT Research Institute, Chicago, Illinois
National Bureau of Standards, Gaithersburg, Maryland
Southwest Research Institute, San Antonio, Texas

Full scale tests have been carried out by several organizations on aircraft fuselages, lavatories, busses, even full buildings, but the results can only be interpreted with reference to the system involved.

3.4 Discussion of Test Methods

Laboratory tests are designed to evaluate some aspect of the fire performance of a material or assembly in a reproducible simulation of some real life situation. This "small-scale" test is useful since it is economical to conduct in comparison to tests on full-scale replications of rooms, corridors, buildings, etc. The major problem, however, is that there is a general lack of correlation between small-scale laboratory test results and real-life large-scale fire behavior.

In this chapter, current tests employed to measure the flammability characteristics of materials will be discussed and in particular, their usefulness to evaluate synthetic polymeric materials will be addressed.

3.4.1 Ignition Tests

Ignition of a combustible material is the first step in any fire and once the fire is started, the igntiion delay times of other materials, coupled with flame spread, affects the rate of fire spread and growth.

Most polymeric materials will ignite and burn if sufficiently heated in the presence of air or other atmosphere containing sufficient oxygen. Consequently, from the point of view of fire safety, it would be desirable to know how long it will take a particular polymer to ignite, or whether or not it will ignite, when exposed to a given oxygen containing environment and a given heating rate.

The thickness and thermal properties of a material are vital in determining the time required to achieve ignition, given a heat flux and environmental condition.

In the case of a thin flammable material (carpet, paneling, etc.) in thermal contact with an underlying material, the thermal properties of the underlying material can strongly influence the ignitability by the degree to which the underlying material acts as a heat sink.

Configuration of the polymeric material can also be of great importance.

The complexity of the ignition phenomenon makes it impossible to devise a single test that will determine the ease of ignition of different materials for a variety of end-use situations.

Another series of ignition tests developed at UL to evaluate the ease of ignition of materials intended for electrical applications involves subjecting the specimens to a high-voltage or high-current arc. Generally these tests are "go-no go" tests, in which the specimens either ignite or do not.

Two recently developed ignition tests have avoided the complexities mentioned above by employing an everyday ignition source, a cigarette, with materials in a specific application, either as a mattress or a covering and padding for upholstered furniture. Despite the simplicity of these tests, once the optimum arrangement and location for placing the ignition source was determined, the results are a good indication of the hazard for these common occurrences.

There seems to be little current development of additional ignition tests. Due to the complexity of the ignition phenomena, it would seem that future ignition tests should be directed towards determining whether or not a given source, e.g., match flame, flaming grease pan in a kitchen, etc., ignited a given item, e.g., drapes, kitchen cabinets, etc. This approach, determining ignitability in a specific instance, could be based on fire statistics data of most probable "ignition source – ignited item" combinations, and would give an accurate measure of the ease of ignition under the specific circumstances of the test. An alternate approach would be to measure some ignition property, e.g., ignition time, in a carefully controlled experiment, recognizing that the quantity measured is not an intrinsic property of the material, and attempt to establish some relationship of this "property" to fire safety in actual situations.

3.4.2 Flame Spread Tests

Unless a person is in intimate contact with the ignited item, a fire is not apt to do much harm until it has grown by spreading some distance from the point of ignition. The rate of flame (and fire) spread is very important in the history of a fire, because it controls the time after ignition when the fire reaches a dangerous size. The "dangerous size" relates to the rate of heat release, to the rate of generation of toxic and smoky products, or to the difficulty of extinguishment. The ability to detect, fight, or escape from the fire depends on the time before the fire reaches a dangerous size and hence, on the spread rate.

A large body of test methods have evolved over the past thirty years designed to measure the rate of flame spread. Many of these tests were developed without allowing for the numerous factors influencing the flame spread rate.

The results from many experimental investigations have been summarized by Friedman, Magee, and McAlevy and indicate that the flame spread velocity is affected by many physical and chemical parameters.

For example, vertical flame spread is a continuously accelerating process; and for small specimens the flame spread rate upward is at least an order of magnitude faster than vertical downward or horizontal flame spread. Also, horizontal flame spread over specimens with exposed edges occurs approximately five times faster than when the edges are not exposed. Raising the initial temperature of polymethylmethacrylate from 25°C to 150°C doubles the horizontal flame spread velocity. These examples show how important an understanding of the factors including flame spread is when either developing or selecting an appropriate flame spread test.

Since heat must travel ahead from the flame to the unignited material, to propagate the flame, certain heat transfer modes must be involved. Thus, tests which presume to establish relative flame spread characteristics of polymeric materials in a definitive manner are valid only for the particular conditions of the test and may give misleading results if extrapolated to other conditions.

Small-scale laboratory screening tests in this category include ASTM D-635, D-1692, D-3014, D-568 and D-1433; all these tests claim usefulness for determining the relative or comparative burning characteristics of materials. Actually they compare the behavior of materials only under the unique conditions of the test. Also, since none of these tests simulates the energy feedback from the surroundings that occurs in real fires, the magnitudes of the spreading flame velocities obtained are unrealistically low.

When a material is used as an interior wall or ceiling finish material, it will affect the fire hazard at the place of use according to the extent to which it permits spread of flame over its surface, contributes fuel to the fire, or generates smoke and toxic gases when burning. In the United States, the Steiner Tunnel Test (ASTM E-84) is the most widely used procedure to measure the potential hazard of room lining materials. While this test reports measurements of fuel contribution, smoke density and rate of flame spread, this test procedure is primarily used to determine flame spread. The test is used in several national model building codes, many local codes, and various regulatory bodies, primarily as a means of limiting the use of combustible interior finish materials in buildings.

The ASTM E-84 test seemed adequate until the development of certain types of synthetic polymeric materials such as low density polyurethane and polystyrene foams, and several types of thermo-plastics. The test method does not provide a satisfactory rating for materials which soften, melt and drip or are of very low density.

Thermoplastic foams rapidly melt in the vicinity of the impinging gas flame and pull away from the roof support. This physical phenomenon prevents flame spread down the tunnel and consequently the material is given a low flame spread rating.

The low density of these foams also results in a misleading indication of fire hazard in this test. Due to a low density and hence a low thermal conductivity, only a thin surface layer of foam is heated by the hot gases in the tunnel. Thus, the rate of production of combustible pyrolysis gases is low and the gases are swept away before burning. In a room fire however, these combustible pyrolysis gases would be confined and eventually burn, augmenting the fire intensity in the room. Thus some materials burn in a very hazardous manner in the real world whereas the test indicates that they should not.

Work on increasing the utility of this test for synthetic polymers, especially foams, is underway.

In a recent series of tests conducted at UL, the flashover characteristics of rooms lined with low density foamed plastics and other common building materials used for interior finishes were investigated. Various ignition sources were employed, and it was concluded that:

> "The flame spread classification of materials developed in the standard 25 foot tunnel test corresponds with the performance of those materials in corridor, corner and vertical-wall full-scale building geometry tests."

This conclusion was based on a criterion for acceptance of the determination of whether the room reached flashover conditions as evidenced by a ceiling maximum temperature greater than 649°C. (1200°F) or if full ceiling involvement occurred.

However, if the time to flashover is employed as the criterion for hazard, then a different conclusion would be reached. Data in the UL report indicates little correlation of flashover time with flame spread classification — in fact, for materials with the same classification, flashover times can vary by an order of magnitude.

The above examples show that the ASTM E-84 test does not consistently measure the hazard from wall and ceiling linings particularly when some new synthetic plastics are employed. Much additional work is required to determine to what extent and for which materials this can adequately measure fire hazard.

Various organizations (e.g., Union Carbide, Monsanto, Forest Products Laboratory) have developed small tunnel tests as a more modest means for evaluating materials than the ASTM E-84 tunnel test.

In the late 50's, NBS developed the radiant panel test with the specific objective of providing a relatively simple and reproducible method for measuring the surface flammability of materials. This test method was recognized by ASTM in 1960 (E-162). It has not been widely used in building code or rating work because E-162 includes a statement that it is not intended for such purposes.

The test recognizes two important factors in characterizing surface flammability: (1) the critical energy flux necessary to propagate the flame and (2) the rate of heat liberated during flame spread.

It seems fair to suggest that use of both the Steiner tunnel test and the radiant-panel method has provided a considerable increase of our understanding of the flammability behavior of solids. However, a need still exists to conduct carefully planned research programs on the relevance of surface flame spread ratings currently being used to determine fire hazards, particularly in the case of synthetic polymeric building finish materials.

Some tests in which flame spread is a factor have been developed to measure a specific hazard. For example, DOC FF 3-71 flammability standard for children's sleepwear is intended to control the hazard of flame spread continuing upward if the fabric is ignited. Similarly, ASTM D-2859 (or DOC FF 1-70) measures the ease with which a small ignition source, simulating for example, a burning ember from a fireplace, could propagate a flame over a carpet or rug. Both these tests form the basis for flammability standards for children's sleepwear, rugs and carpets respectively.

Another test specifically designed to measure surface flammability of a specific item, i.e., roof coverings, in the configuration and under exposure conditions which might be typical of a real-life fire, is ASTM E-108. This test seems to be a reasonable approach to classifying roof coverings.

In summary, the surface flammability characteristics of materials are important and should be evaluated. However, since the flame spread process is strongly controlled by the magnitude and relative importance of the heat transfer modes involved, small-scale tests designed to provide relative comparisons of hazards have very limited meaning since they fail to model thermal energy feedback correctly.

On the other hand, small-scale tests developed to simulate a specific surface flammability hazard (e.g., carpet test) can be quite effective in controlling that particular hazard.

Much work must be done to correlate surface flammability test ratings to full-scale fire behavior hazard. However, it is highly unlikely that any one flame spread test can predict the surface flammability hazard of all materials in all situations.

3.4.3 Rate of Heat Release Tests

In recent years, there has been growing support of the concept that the rate at which heat is released during burning is an important criterion for evaluating the fire hazard from a particular material.

Since the rate of heat release is thought to be most important during the "steady" burning period following flame spread, it is a measure distinct from ignitability or surface flame spread potential, and is considered to be of greatest significance in the stage of fire growth preceding flashover.

The rate of heat release from initially ignited material(s) has a significant influence on local fire intensity and hence on the subsequent development of a fire. Thus a basis for the importance of the rate of heat release concept is readily established.

The concept of rate of heat release had its origins in the work of Steiner when he described the measurement of "fuel contributed" along with flame spread rate in his then newly developed tunnel test (ASTM E-84). Robertson, Gross, and Loftuz in what is now ASTM E-162, combined a flame spread rate with a factor involving the rate of heat generated by the material under test to determine a flame spread index for that material. Both of these tests are essentially tests for surface flame spread, and when evaluating heat release, these tests report total heat released, and only provide information relative to red oak (ASTM E-84) or hardboard (ASTM E-162).

The first instrument designed specifically to determine rate of heat release was the FM Construction Materials Calorimeter developed at Factory Mutual Laboratories in 1959. Recently, instruments have been developed at the National Bureau of Standards (NBS), Ohio State University (OSU), Stanford Research Institute (SRI) and Forest Products Laboratory (FPL) to measure the rate at which heat is released from a burning material.

In a typical heat release rate calorimeter (HRRC), a sample of material, of known physical and chemical composition, is exposed to a controlled air flow and an external radiant heat flux simultaneously. The back surface of the specimen is either insulated or water-cooled. In the latter case, the temperature rise of the water stream is used to calculate the heat "released" through the specimen back surface.

Once the specimen is ignited, heat is released as a function of time. This released heat may be measured directly or by substitution, by either operating the calorimeter in an isothermal or adiabatic mode. In the substitution method, the flue gas time-temperature curve is reproduced by burning a gas, e.g., propane, to make up the difference between the test and reference (inert) samples. The rate of heat release of the test sample is thus obtained from the rate of consumption of propane during the substitution run. During adiabatic operation of the HRRC, the rate of heat release can be calculated directly from the rate of temperature rise in the products of combustion.

Since the rate of heat release is not a fundamental physiocochemical property of a material, one finds that different HRRC's, employing different "fire conditions," yield different measures of rate of heat release.

Each HRRC has one or more shortcomings. For example, both the FM and FPL HRRC's operate on the substitution principle and hence, they require two runs to evaluate each sample. The FM calorimeter also employs an exposure of the sample which corresponds roughly to the standard ASTM E-119 time-temperature curve compressed into a much shorter time, however it suffers since it only allows one fire exposure. The primary disadvantages of the NBS instrument are small sample size, limitations in evaluating an assembly of materials in a practical configuration, and the need to use an after-burner. Both the OSU and FPL calorimeters are limited to the low and moderate heat flux ranges. The OSU instrument also suffers in that the rate of heat release due to flame spreading over the surface cannot be uncoupled from the rate of heat release from those portions over which the flame spreading process is completed. Also, since the OSU instrument operates adiabatically, with a direct measurement, the continuous application of a pilot flame at the specimen bottom edge will influence the heat release rates. The SRI instrument was specifically designed to imitate the NBS calorimeter, maintaining the high heat flux capability but increasing the size so that much larger specimens could be accommodated. It appears that attempts to compare data derived from any of the five calorimeters with those from any other of the calorimeters will be unsuccessful in large measure because of the overriding dissimilarities in operation.

Current HRRC's are not restricted to any specific material and thus are suitable for the determination of rate of heat release from polymeric materials.

The development of a test method generally includes three stages: the growth of a concept, the design and development of an instrument to measure a quantity based on the concept, and the application of the measurements to a specific situation. Test methods for rate of heat release are presently at stage two. After completion of this step criteria will be developed, based on rate of heat release tests, to improve the fire safety of materials in use. Meanwhile, more data are needed to fully evaluate the performance characteristics and reproducibility of the currently available instruments. Current HRRC's differ significantly in their designs and capabilities, and most are still being modified. Much more round-robin testing should be done to determine the accuracy and precision of each instrument.

A recently developed test, which in essence measures rate of heat release from wearing apparel, is known as the "Mushroom Apparel Flammability Test" (MAFT). This test has been proposed as the basis for a new Federal Standard for flammability of wearing apparel. Fabric specimens are tested in a cylindrical configuration (which simulates a garment). Pass-fail criteria are based on (a) time to ignite with a specified gas flame, and (b) on heat transferred to sensors in the apparatus.

A fabric classification scheme, based on a maximum heat transfer rate and minimum ignition time, is proposed as part of the standard, and suggests that fabrics which transfer little heat to the inside of the cylindrical configuration (Class I) could be used for all garments. Use of fabrics which transfer larger amounts of heat would be restricted as to garment type and style. The conditions of the test are intended to simulate realistic situations.

3.4.4 Tests for Oxygen Requirements

Since its introduction in 1966, the Oxygen Index Test has achieved wide popularity as a means of characterizing the flammability of materials. Its advantages include simplicity of equipment and ease of operation, good reproducibility, good differentiation of materials, versatility, and the use of small, easily prepared test specimens. The standard version of the Oxygen Index Test, ASTM D2863, is applicable to plastics including solid, cellular, and film forms. A similar procedure has been used extensively for textile materials.

In the standard test, the sample is ignited at the top and burns downward with a small candle-like laminar flame. Energy from the flame is dissipated to the cool surroundings and there is little energy feed-back to support the combustion of the sample. This is in sharp contrast with the hot turbulent environment characteristic of most real fires. For these reasons, the test is of little value in characterizing the behavior of materials under conditions encountered in real fires.

Thermoplastic materials present problems in the standard test because of melting and dripping. The melt carries heat away from the combustion zone so a higher oxygen concentration is needed to maintain the heat balance at the burning surface. To avoid this problem, wicks or cup type sample holders have been used to hold the melt in place. Good correlations of oxygen index with polymer viscosity have been observed. This may be important in studying the effects of additives on polymer flammability characteristics. Additives which lower the melt viscosity may increase the oxygen index, giving a false indication of flame retardant action which may not be supported by other tests.

The oxygen index decreases with an increase in the initial sample temperature since less energy feedback is needed to maintain the burning surface of the sample at the required temperature. This can be a cause of error when the apparatus becomes heated from successive tests.

The oxygen *sensitivity,* the rate of change of O! with temperature in the atmospheric oxygen concentration region, may be another useful measure of flammability.

Numerous attempts have been made to provide a theoretical interpretation of the oxygen index test, but the results to date have been unsatisfactory. While the sample is burning, energy feedback from the flame to the burning surface maintains the surface temperature at a level where pyrolysis of the polymer supplies gaseous fuel to form a combustible mixture with the stream of oxygen plus nitrogen. As the oxygen concentration is decreased (and the nitrogen concentration increased) the flame temperature will decrease, reducing heat feedback and the supply of fuel to the flame zone. At a critical oxygen level (the oxygen index) a sudden transition from active burning to extinction occurs.

Attempts have been made to apply the oxygen index test to liquids. From a practical standpoint, the standard test configuration suffers from the disadvantage that the sample holder provides a large heat sink with poorly defined heat transfer characteristics. The measured value of the oxygen index under these conditions is more a function of the design of the apparatus than an intrinsic property of the liquid. The study of liquids is attractive from the theoretical point of view since the vaporization process is simpler and better understood than pyrolysis of polymers.

The difficulties encountered in these experiments with liquids help to show why the oxygen index test is successful with solid polymers. Most solid polymers have low and similar values of thermal diffusivity. The temperature of the burning surface is high and relatively constant for all organic polymers since it depends on the strength of the carbon-carbon bond rather than on the vaporization of small molecules. When an elongated polymer sample is burned at its upper end, heat loss from the hot burning surface into the sample and sample holder is small and relatively independent of minor variations in composition and sample size. The heat balance at the burning surface depends primarily on the chemical properties of the polymer and the composition of the atmosphere and is relatively independent of apparatus parameters. Thus the test provides a convenient way of comparing the relative flammability of similar polymers compositions under closely controlled laboratory conditions. Recourse must be had to test methods providing a much more severe fire exposure in order to predict the performance of a material under conditions likely to be encountered in real fires.

3.4.5 Extinguishment Test Methods, Specifications, and Standards

3.4.5.1 Portable Extinguishers

In terms of extinguishment, combustibles are classified by NFPA in four categories:

Class A: "ordinary combustible materials, such as wood, cloth, paper, rubber, *and many plastics*"[1]
Class B: "flammable liquids, gases, and greases"
Class C: "energized electrical equipment"
Class D: "combustible metals, such as magnesium, titanium, zirconium, sodium, and potassium."

More than half the weight of synthetic polymers currently used consists of materials which fuse on application of heat to produce a low-viscosity melt. On the other hand, other polymers produce a rather viscous melt which tends to stay in place rather than flow, and yet others char rather than melt. Furthermore, different solids of the same size and shape which are fully ignited will burn at different rates. Thus, polyethylene burns at 1/3 the rate of polystyrene. Char-forming materials burn even slower relative to polystyrene, which is the fastest burning unfoamed solid of more than a dozen tested. (See Vol. 2 for details). On the other hand, liquid styrene monomer burns more than three times as fast as solid polystyrene, under certain conditions.

There is no standardized way of rating ease of extinguishment which takes into account these factors of different burning rate and different tendency to form a fluid melt, except in the sense of a Class A versus Class B classification. Plastics are generally taken to be Class A combustibles. Portable (hand-held) extinguishers intended for Class A fires are tested by approval agencies (UL, Factory Mutual) for their ability to extinguish a specified stack of hardwood pallets, after a specified preburn time. Also, excelsior (wood shavings) is sometimes used to represent a Class A combustible in these tests.

Class B combustibles are simulated by a square steel pan containing n-heptane. The larger the capacity of the extinguisher, the larger the area of test fire which must be extinguished in an approval test.

It is probable that a fire involving a polymer which forms a fluid melt can be controlled by an extinguisher intended for Class B fires. Application of water to such a fire would probably cause spattering and temporary augmentation of burning rate, but continued application of water would cool the polymer to a non-burning state. Note that, since the heat of gasification of a molten polymer is several times as high as that of an organic liquid such as heptane or benzene, and the flash point of the polymer is also much higher, molten polymer fires would be considerably less difficult to extinguish than heptane fires. On the other hand, if the burning polymer is a smoldering cellular material, the fire being deep-seated, attack with a Class A extinguisher (rather than Class B) would appear to be more appropriate.

The foregoing paragraph is a summary of the fire protection engineer's knowledge about fighting polymer fires with hand-held extinguishers. There are no stan-

---

[1] Emphasis supplied.

dardized test methods to determine the best extinguisher for a particular burning polymer. Such test methods might be useful, but the value would be limited severely by two factors. First, a given portable extinguisher may be used to fight fires of a wide variety of materials, making optimization difficult. Second, the difficulty of fighting a fire depends more on the size of the fire than on the type of combustible. Hence the need for better test methods in this area would appear to have low priority.

3.4.5.2 Automatic Sprinklers

Automatic sprinklers to protect buildings are specified on the basis of rate of delivered water volume per unit area ("water density") (generally between 0.1 and 0.6 gallons/min-square foot) and the maximum foreseeable water demand (sometimes as large as several thousand gallons/min). The cost of installing the sprinkler system is highly dependent on these design parameters.

The "water density" required to control a fire depends on the type, quantity, and arrangement of combustible present. Polymeric materials in some cases, because of high burning rates, may present especially severe challenges to sprinkler systems. However, in general, fire loss to property, especially commercial and industry property, has been tremendously reduced when automatic sprinklers have been present. Insurance statistics show that the expected loss due to fire in sprinklered properties is about $2 per year per $10,000 value. Data suggest that it is about an order of magnitude higher for unsprinklered property.

Private residences are virtually all unsprinklered, because of cost. A major proportion of hotels, school dormitories, etc. are sprinklered. A greater proportion of nursing homes, restaurants, etc. are sprinklered, depending on local ordinances.

Standards for acceptance sprinkler protection of commercial and industrial properties have been obtained and are continually being refined by insurance companies, e.g., Factory Mutual, partially by full-scale fire tests and partially from loss experience. In a few cases, modeling experiments have been helpful.

To illustrate a difference in burning of wood versus expanded polystyrene under sprinklers, the following data from Factory Mutual are quoted:

Common Conditions: Pallets made of either wood or expanded polystyrene, stacked six feet high (9 stacks in square arrangement), with sprinkler water density of 0.3 gal/min-square foot, top of array 9 feet below ceiling, 165°F sprinkler links spaced 10-feet by 10-feet on ceiling.

Results – Wood: Seven sprinklers opened; maximum ceiling temperature above fire was 296°C (565°F). The fire spread was controlled.

Results – Polystyrene: 59 sprinklers opened; maximum ceiling temperature above fire was 1150°C (2100°F). Fire burned intensely until melting caused collapse of the pile.

For very intense fires, especially under high ceilings, the upward velocity of the fire plume prevents penetration of water droplets. By modifying sprinkler design to increase production of large drop sizes, penetration can be dramatically improved. By operating sprinklers above a fire plume maintained and closely controlled by feeding gaseous fuel to a ring of burners, and measuring water arrival at floor level in the center of the ring, one can measure penetration efficiency as a function of sprinkler parameters. Factory Mutual is currently attempting to develop sprinkler standards based on these considerations.

One of the most severe fire challenges known, other than flammable liquids or gases, is that of plastic commodities which are stored in factories, warehouses or retail outlets, in configurations reaching to substantial heights.

It may be remarked that it is possible to store polymeric materials safely at greater heights than those permitted under ceiling sprinklers by installing additional sprinklers at intermediate levels, preferably below horizontal barriers. Alternatively, a development program is underway at Factory Mutual to use unconventional sprinkler heads of greater penetrating power at ceiling level; this would permit relaxation of either storage height requirement or total water demand.

In contrast to industrial fires under high ceilings, there is very little experience on the minimum sprinkler requirement for a fire in a normal size residential room. There is reason to believe that very little water may suffice to control such a fire. However, if it is necessary to spray the water directly on the burning object, it is difficult to locate a single sprinkler head so that the spray will impinge on every potential combustible in the room; if several sprinkler heads are needed per room the installation cost is high. Data has been developed relevant to another concept – i.e., spraying the sprinkler water into the fire gases and onto the hot ceiling and walls so as to generate steam which may extinguish the fire by inerting. Less than 0.5 gal/min-square foot may be adequate. Much testing is needed with realistically furnished rooms before this concept can be standardized.

The action of a sprinkler to control a fire is in part by inerting (both by steam and by entrained combustion products), in part by direct wetting of the pyrolyzing solid, in part by pre-wetting not-yet-ignited materials, in part by cooling the fire gases – thus reducing radiative transfer to combustibles, and in part by creating a fog which itself blocks radiative transfer. In view of these multiple effects, there seems to be little chance that a small scale test method evaluating any one of these effects will suffice to determine minimum water application rate for a given polymeric commodity.

3.4.6 Tests for Smoke Evolution

Tests for smoke[2] evolution are important to fire safety because the presence of smoke affects the ability of occupants to escape from a burning structure and the ability of firefighters to control and extinguish the fire. Most existing smoke test methods are concerned with vision obscuration; therefore they seek to measure

---

[2] In the following discussion smoke is artifically distinguished from the toxic products of combustion.

smoke density by either optical or gravimetric techniques. No smoke method presently in use addresses an important aspect, the lachrymatory and irritant characteristics of smoke which can obscure vision even more effectively than optical density in the atmosphere.

The NBS smoke density chamber is a most widely used apparatus for measuring smoke density. This represents the best available means for evaluating the relative smoke producing characteristics of a material in terms of obscuration.

The NBS smoke density test method, also designated NFPA 258, exposes a specimen three inches square to a radiant heat flux of 2.5 W/cm^2. The smoke evolved is accumulated in an enclosure whose volume is 18 $ft.^3$. Optical density is measured along a vertical light path three feet long. The results are expressed in values of specific optical density (Ds), and the maximum value is designated as Dm.

It is when use of the test method is expanded from its basic function of obtaining comparable data on materials that some aspects can become questionable. As examples, consider the following (fuller details are in Volume 2):

1. The maximum value of specific optical density, Dm, is corrected by subtracting the optical density due to soot deposited on the optical system. This procedure favors heavy soot-depositing materials such as polystyrene to the possible detriment of fire safety. The FAA proposal to use Ds values at 1.5 and 4.0 minutes is to be commended because it avoids the correction step. Fortunately for public safety, the Dm values for heavy soot-depositing materials are so high even with the correction that these materials are still recognized as hazardous.

2. Specific optical density is a useful scale for expressing relative smoke density. Its use in extrapolating values to larger enclosures, however, can be questioned.

3. The vertical specimen position permit materials which melt to exhibit unrealistically low values because the molten material either is subject to less heat in the specimen holder trough or overflows the trough and escapes exposure entirely.

The NBS chamber offers the advantages of considerable versatility, such as the addition of controlled ventilation, continuous weight monitoring, and analytical and bioassay devices for measuring toxicity. It offers one-dimensional heat flux, which is essential for evaluating composite structures which are expected to encounter heat flux on one side only.

The ASTM D 2843 test is a useful laboratory screening test, and can distinguish gross differences between materials at substantially less cost than with NBS smoke chamber.

The principal advantage of the ASTM E 84 test in regards to smoke is that a great many materials have been tested by this method in the absence of other widely used tests. However, it is desirable that smoke evolution be measured by a test specifically designed for the purpose.

3.4.7 Tests for Toxic Gas Evolution

Tests for toxic gas evolution use one (or both) of two approaches: identification and analysis of the chemical compounds in the combustion gases, and exposure of laboratory animals to the combustion gases.

The analytical approach is based on the philosophy that if only a limited number of chemical compounds are responsible for essentially all the toxic effects, then analysis of these particular compounds would permit prediction of essentially all the toxic effects which could be expected.

The bioassay approach is based on the philosophy that the response of a laboratory animal to the combustion gases in effect integrates all the toxic effects, and therefore a material which is more toxic to an animal than another would likewise be reasonably expected to be more toxic to man, regardless of whether it is supposed or predicted to be or not.

Both approaches have common problems: loss of toxicants or change in toxicants between the source of toxicants and the sensor of toxicants.

It is generally accepted that carbon monoxide is the principal toxicant in most fires. The toxicity goals of the Boeing Commercial Airplane Company specify, not only CO, but also HCN, HF, HCl, SO_2 and NO_2. If O_2 is added to measure depletion and CO_2 is added to measure hyperventilation, the total is brought to eight compounds. If analysis for these eight compounds can result in adequate prediction of relative toxicity, then a GC-MS-computer system becomes unnecessary.

The problem of interferences is significant when using highly sophisticated instruments. In some cases, traditional wet chemical analysis is still the reference method despite the use of expensive instruments.

Physical collection of gaseous toxicants for analysis can also present difficulties, and various trapping systems have been devised to perform this task.

In most cases, analysis for toxicants must be performed immediately after the test, because some toxicants such as hydrogen cyanide are developed or lost with time through various mechanisms.

The analytical approach, if adequately validated, using only a limited number of toxicants to be analyzed, offers significant advantages in terms of simplicity, logistics and other problems involved in bioassay screening.

The widely-distributed NBS smoke density chamber appears to be widely used for analytical toxicity studies because gas analysis can be done in conjunction with smoke density.

Perhaps the basic weakness of the analytical approach is the assumption that knowledge of the quantities of various toxicants present can lead to a satisfactory prediction of their combined toxic effects. This overlooks synergistic effects that may be significant.

It is on this issue that the bioassay approach offers the major advantage of simplicity. In contrast to the analytical approach that calls for elaborate techniques for analysis followed by hopeful prediction of toxic effects on the basis of incomplete data for pure gases, the bioassay technique directly measures the effect of the combustion products on a biological system, and bypasses all the intervening uncertainties. Since human beings cannot be used in such experiments, a choice of test animal must be made, and mice and rats are the species of choice in most tests. There are differences of opinion as to whether each species, and each strain of each species, is more suitable for predicting human response, but the only clear advantage of rats over mice is that the rat, being a larger animal, provides a more ready supply for blood analysis. If blood analysis is not a requirement then the mouse offers the advantages of smaller size, lower cost in both purchase and maintenance, lower oxygen consumption per animal, and the willingness to run in an exercise wheel without motor drive.

The toxicity tests utilizing the bioassay approach vary considerably in both apparatus and procedure. The smaller the chamber, the greater the effect of oxygen consumption by the test animals over a given period of time; this limits the number and size of the test animals.

On the other hand, the larger the chamber, the greater the quantity of material required to achieve a given concentration of toxicant, and the greater the probability of variations in gas and aerosol composition between different locations in the chamber; there is also proportionally greater surface for condensation and adsorption of toxicants, and a proportionally smaller fraction of the volume that the test animals actually have access to.

The available tests differ as to whether the test animals are in the same chamber as the material being tested. The major reason for separating the combustion chamber by a given distance from the animal chamber is to cool the gases sufficiently and minimize direct radiation so that the animals are not exposed to extreme temperatures.

Toxicity tests also differ as to whether a fixed temperature or rising temperature history is used. Many materials produce more toxic effluents at some temperatures than at others, and the selection of temperature levels would unavoidably favor some materials over others. The rising temperature history exposes the material to the entire range of temperatures which it must undergo on the way to the upper limit temperature, and is, therefore, more realistic than an extremely rapid rise to the designated temperature.

There are differences between tests as to the amount of air flow used. Some tests are entirely closed-system tests, with neither inflow of fresh air nor bleeding off of toxicants by displacement. The results are independent of flow rate and involve no loss of toxicants, but the lack of a fresh air supply does not alleviate oxygen depletion. Some tests use closed systems but add continuous recirculation. Other tests use the flow-through technique with a continuous stream of air; this technique offers the lowest oxygen depletion and the highest level of oxidation during pyrolysis.

Most toxicity tests use pyrolysis, and a minority use flaming combustion; among the latter are the NBS smoke density chamber in flaming mode, and some tests used at DuPont, Harvard, University of Michigan, and University of Tennessee. Flaming combustion introduces more variables and aggravates the problems of oxygen depletion and heat production as regards the test animals.

The apparatus should be considered separately from the procedure. The simplest procedure is to perform the test and count the number of dead animals at the end, and report percent mortality. A second approach is to continue testing, varying the quantity of sample, until the amount that produces death in 50 percent of the test animals is determined (LD_{50} or LC_{50}). This provides a continuous ranking of materials, based on the concentration needed to produce a certain effect. A third approach is to use a fixed weight of sample and record time to incapacitation and time to death. This provides continuous ranking of materials, based on the time required to produce a response from a fixed sample quantity. The rankings using these two bases are not necessarily identical, and since both have some validity a combination of the two is desirable.

Beyond these approaches, the changes become technique rather than philosophy. With many systems, the apparatus provides poor visibility of the test animals. To measure incapacitation, exercise wheels and shock reflexes have been used. To measure death, respiration chamber respirometers, EKG, and EEG have been used. All involve some form of restraint and the loss of observation is obtainable only from freely moving animals. Exercise wheels increase oxygen consumption by the test animals and the exertions add to stress. The other techniques involve restraint to the point of immobilizing the animal, and some require prior training. With sophisticated instrumentation comes increased vulnerability to electronic malfunctions.

It is worth noting that despite all the differences in apparatus and procedure, the relative toxicity rankings of many materials tend to be the same in most tests which have similar conditions of pyrolysis or combustion, although actual values may vary somewhat.

The simplest screening test is the USAF/NASA toxicity screening test method. The apparatus consists of a horizontal tube furnace, a quartz pyrolysis tube, and a liter hemispherical chamber of patented NASA design. Four Swiss albino male mice are used in each test. Two or three tests are performed, to provide replication between animals and between tests. A single operator observes the freely moving mice for times to incapacitation and death. Selected operational cycles have resulted in almost all materials giving time to death of between 5 and 30 minutes. At least 120 materials have been reported thus far.

The next stage of sophistication is represented by the FAA/CAMI test method. The apparatus consists of a horizontal tube furnace, a quartz pyrolysis tube, a rectangular box with exercise wheel, and a recirculation system, all with a system volume of 12.6 liters. Three Sprague-Dawley male rats are used in each test, and two or three tests are performed. Pure oxygen is periodically introduced to maintain oxygen level at 21 percent. Two operators observe the rats for time to incapacitation and death. At least 15 materials have been reported thus far.

With increasing sophistication it becomes difficult to identify any particular tests as characteristic of succeeding stages.

The NBS smoke density chamber provides poor visibility, even with an animal module accessory, but the 510 liter volume of gas available for analysis is an advantage when both analytical and bioassay approaches are used simultaneously.

For those who wish to pursue the analytical approach with minimum investment, the NBS smoke density chamber with provisions for analyzing eight or fewer gases appears adequate.

For those who wish to pursue the bioassay approach with minimum investment, the USAF/NASA screening test appears appropriate.

For a more sophisticated bioassay method, the FAA/CAMI method seems appropriate. The larger system volume provides more gas for analysis, and the motor drive exercise wheels provide a more definitive measure of incapacitation than the free movements of the mice in the USAF/NASA method.

3.4.8 Tests for Fire Endurance

The chain of events that leads to a serious fire can be broken in many places. Often the probability of a fire reaching flashover is very small, but a sequence of improbable events can lead to flashover in even the best protected areas. The realization that the worst eventuality is not impossible must lead to an evaluation of how a building would respond to a sustained high-intensity fire. This structural response is termed fire resistance or fire endurance and is measured by the ASTM E-119 Fire Test.

A number of factors, such as ventilation, fuel load, and thermal properties of the structure, have a controlling influence on fire intensity at any instant.

The earliest attempts to test the fire resistance of different structures occurred in the last two decades of the 19th century and essentially consisted of some *ad hoc* furnace tests in which the specimen was exposed to a constant temperature for a given time period.

In 1918, an ASTM committee issued the first edition of C-19, the predecessor of the current E-119 standard. The C-19 standard was based on an exposure of the test specimen to a time-temperature curve which was believed to approximate the course of fire in the heavily timbered structures of that day. This curve has been essentially unchanged and, to this day, it closely resembles the curves used by most countries for the post-flashover evaluation of structures.

After the initial adoption of the standard post-flashover fire test, there arose some interest in checking the validity of the assumptions that underlay the committee work. The most significant research was attempted by Simon Ingberg at the U.S. National Bureau of Standards. Starting in 1922 and continuing through the 1940's, he conducted numerous research programs aimed at better characterizing the post-flashover fire.

Ingberg's experimental work probed two aspects of describing the post-flashover fire burnout tests and fuel load surveys. The burnout experiments were conducted over a period of years in three different test buildings and encompassed both residential and office occupancies.

One feature of these tests was Ingberg's attempt to obtain the worst possible fire conditions by controlling ventilation. He normally adjusted swing shutters on the windows of his burnout chambers to achieve his purpose.

From the earliest tests it was apparent that these fires did not reproduce the ASTM time-temperature curve very well. A single standardized curve was so appealing, however, that Ingberg devised a way of molding reality to fit the curve. The stratagem that he invented was to define fire "severity" which was set equal to the integral under the time-temperature curve, above a baseline of either 150°C or 300°C. As can readily be proved, this equal-area "severity" has no physical justification. It did, however, conveniently reduce the fire description from a two-variable into a single-variable problem, so that different fires could be directly compared. The standard curve was thus saved – any actual fire could be defined to have an equivalent duration on the standard test.

To make these results useful, the expected fuel load has to be known. Fuel load surveys were conducted by NBS at several times, and are still continuing. This information has not been reflected in many building codes.

The most important aspect of the ASTM E-119 fire test, and all simulated post-flashover tests, is that it purports to allow the evaluation of performance of a structural element under a standard fire exposure. It is not used to evaluate whether the structural element will contribute to the propagation of a fire within a compartment, but rather the load carrying capacity or the fire containment performance of the element.

Maintenance of a positive pressure within the furnace would probably improve the ASTM E-119 test method, but at this time there are no requirements for measuring and/or regulating the relative pressure between the inside and outside of the furance. Under actual fire conditions in fully involved compartments, a positive pressure exists in the upper two-thirds of the compartment. It is particularly important that E-119 tests be conducted with a positive pressure on at least the upper two-thirds of wall specimens if there are openings in the wall such as pipe and conduit penetration and doors (doors are tested by ASTM E-152, the special version of E-119 that deals with doors).

3.5 Test Development Based on Fire Dynamics

3.5.1 Introduction

If the material were expected to be exposed to fire risk in only one precisely defined set of circumstances (size, orientation, type of ignition source, method of applying ignition source, ventilation conditions, environmental conditions, etc.), it would be obvious how to test it for fire hazard. One must evaluate a series of candidate materials under a particular set of circumstances and note which materials are satisfactory or unsatisfactory. One would then evaluate smaller samples of the same series of materials by a proposed test method, which would become validated for future use if the results of the two procedures correlated.

Frequently, this idealized approach is not directly applicable because the material of interest may be exposed to fire risk in a wide variety of ways rather than in a simple well-specified set of circumstances.

As a result, many fire test methods are considered to be highly unreliable indicators of hazard. For many other cases, no appropriate test methods exist. Tests are particularly weak in regard to fire spread and growth, as well as noxious gas and smoke production.

If the scientific studies of combustion were sufficiently advanced, one might hope for the development of theoretical links between physical and chemical phenomena involved in test methods and the corresponding phenomena involved in real fires. However, rather limited progress has been made in untangling the complexities of fire behavior and at this time most fire situations cannot be analyzed in fundamental terms. Research, involving highly idealized materials and geometries, is progressing on a limited scale. However, much more effort is needed before we can expect major contributions by combustion scientists to more relevant test methods.

3.5.2 The Scope of Fire Dynamics

Fire dynamics involves quantitative descriptions of fire phenomena on a scientific basis. An understanding of these phenomena may be obtained on any one of four levels; see Section 5.2.2.

3.5.3 Current State-of-the-Art of Critical Fire Dynamics Elements

3.5.3.1 Introduction

A discussion of recent research is found in Section 5.2.3 ff.

To illustrate the current state of the art several samples of fire dynamics studies and their importance to test method development will be discussed.

3.5.3.2 Burning Rate of Plastic Slab

Once fully ignited, a thick plastic slab burns at a rate which is believed to be essentially independent of all chemical kinetic parameters, except those that might affect the luminosity of the flame.

For the case of a small or nonluminous flame, radiation is negligible, and the burning rate is controlled by an energy balance at the surface, in which heat is convected from the flame gases to the surface and absorbed by conduction into the interior, endothermic heat-up, depolymerization, and gasification of the plastic. While the depolymerization reactions obey the laws of chemical kinetics, the kinetics have very little influence on the rate of gasification because the surface assumes a sufficiently high temperature to permit the reactions to occur just fast enough to satisfy the energy balance.

When the specimen has dimensions of about 10 cm or more, the flame will be turbulent (instead of laminar) and large enough so that radiation from the flame to the surface becomes significant. In this case, neither the theory of turbulence nor the theory of flame radiation is sufficiently advanced to permit quantitative calculation of burning rate from first principles, as contrasted with the preceding laminar case.

For example, a horizontal slab of polystyrene, 18 cm square, burning as a pool, is consumed nearly twice as fast (mass per unit area) as an identical slab of polyoxymethylene. In contrast, when a small specimen of polystyrene rod is burned in an apparatus with an opposed air jet, it is consumed at less than half as fast as polyoxymethylene. When burning as a large turbulent pool, polystyrene has a flame which is highly luminous, whereas polyoxymethylene flame is blue. In small opposed-jet burners, both flames are blue.

Additionaly, when a large polymethyl methacrylate slab is burned vertically (on one side) instead of horizontally, the burning rate is found to be lower.

To ascertain the effect of incident radiant energy, the rate of increase in burning rate for burning plastics, with incident radiant energy (supplied by electrical radiant panels) has been measured. For example, a radiant flux of only 1.3 watts/cm^2 increased the steady burning rate of a vertical polymethyl methacrylate slab by a factor of 2.5. Large fires emit radiant fluxes at least five times greater than the value of the example.

Burning rates of thick slabs of wood or charring plastics are much lower than those of melting plastics, for two main reasons: 1) the char layer acts as insulation between the flame and the virgin fuel; 2) the re-radiation of heat from the surface is much higher.

These recently acquired insights into the role of radiation in burning, give promise for a better understanding of these phenomena.

3.5.3.3 Burning in an Enclosure

A wood crib in the open burns at a reproducible rate, which may be correlated with the geometrical parameters of the crib. When such a crib is burned in a non-combustible enclosure with minimal ventilation, the burning rate is greatly reduced, and the proportion of carbon monoxide in the combustion products is greatly increased. However, if the test is repeated in an enclosure that is very well ventilated, the burning rate is considerably higher (up to 70% more) than the burning rate for the same crib in the open.

In a more dramatic experiment, a horizontal slab of polymethyl methacrylate, 30-inches square, was burned under a ceiling four feet high, with adequate ventilation. The slab burned for a short initial period at a rate corresponding to burning in the open, but then rapidly accelerated to a rate about four times as great, as radiant feedback built up.

Since the burning of materials in compartments is common to most fires, further development of the laws governing such burning is essential. And flammability test methods designed to measure the hazard contribution by individual items in a compartment, e.g., an upholstered chair, must account for the effect of the compartment of burning behavior.

3.5.4 Applications of Fire Dynamics to Test Method Development (also see Section 5.2.4)

In general, our knowledge of fire dynamics is still rudimentary. Most tests were developed when such knowledge was even less developed, or at times when enormous pressure existed to develop test methods in a matter of months, permitting no time for realistic fire dynamics approach.

However, in a recent case, fire dynamics considerations were used in test development for fire spread down a corridor with flammable floor covering. Corridor fire experiments were performed; the radiation from the ceiling, as well as from the burning gases just below the ceiling, was identified as critical in promoting a "flashover condition" over the floor covering.

Another approach to test method development is the use of fire modeling. A physical model of a fire situation is reduced in scale while maintaining geometric similarity and reproducing the important chemical and thermodynamic parameters of the materials involved. Progress in model development has been very slow.

The difficulties of modeling a fire by reducing scale arise in several ways: 1) very small fires are laminar while larger fires are turbulent; 2) the ratio of buoyancy forces to viscous forces in the convective flow of fire gases is size-dependent; 3) the radiant emission and self-absorption of the flame are size-dependent; and 4) the gas-phase time scale in the fire is shorter for small than for large fires.

However, it is anticipated that valid modeling procedures may be developed for at least some aspects of fire behavior as the understanding of fire dynamics relevant to situations of interest improves.

Some progress has been made by a pressure-modeling technique based on the principle that the Grashof number is invariant if the product of pressure squared and size cubed is held constant. (The Grashof number is a dimensionless ratio of buoyancy forces to viscous forces in the hot gases around a fire.) Pressure-modeling has been shown to be valid for diffusion-controlled burning as long as radiation either varies with pressure in certain prescribed ways or is negligible. It has worked over a tenfold specimen size range for turbulent burning of polymethyl methacrylate and wood objects in several shapes. Future progress in the modeling of fire situations will be difficult to achieve without a better understanding of the underlying controlling mechanisms of fire.

3.6 Criteria, Specifications, Standards, Codes

3.6.1 Purpose

This section addresses the methods whereby certain documents of concern to the general subject of the fire safety aspects of polymeric materials are used for protecting people against fire and reducing fire loss.

3.6.2 Word Usage

To avoid ambiguity, we use here the definitions of section 3.2.

3.6.3 Origin of Documents

It is apparent that documents are generated by those who need or will use them. A buyer may generate a specification to bind a supplier, a seller may generate a specification to indicate what he can or will supply. Regulators prepare enforcement documents. Professionals use documents to transmit advanced technology.

3.6.3.1 Federal Documents

Federal documents of interest fall into three classes: (1) procurement documents, (2) guideline documents, (3) regulatory documents. The first group sets the basis for competition among suppliers to the Federal Government; the others are self-explanatory.

3.6.3.2 State and Local Documents

Procurement documents are the principal standards activities in most states; they are associated with purchasing activities. Specifications and standards are drafted by purchasing agents. When the subject is reorganized as having a substantial technical content, engineers are sometimes brought into the drafting of the specifications; at other times, a specification from another jurisdiction may be copied.

3.6.3.3 Codes

Codes are mandatory standards. They are used for regulations of many aspects of life. For purposes of this section, we list building codes, state and local fire laws, and fire prevention codes. The National Fire Prevention Association is a principal source of consensus fire standards and codes which have been woven into the body of law by the various levels of government. More than 200 separate standards and codes have been published by the NFPA; they are codified annually in the volumes of the National Fire Codes. These NFPA codes serve as the basis for regulatory standards. The National Building Code is a model code that contains regulations for life safety in buildings and for the fire protection of buildings. A companion document, Fire Protection Code, is also available. These codes only require enactment to become law.

3.6.3.4 Voluntary Standards[3]

Some 116 technical and nontechnical, trade, and other professional organizations have been identified as generating voluntary standards in the area of safety. Of these, perhaps a third relate to fire safety. These organizations cover the entire

[3]See also NMAB-330, Materials and Process Specifications and Standards, National Materials Advisory Board, National Research Council, Washington, D.C. 1977.

spectrum of scope, proportion of activities devoted to standards work, participation in international programs, certification work, degree of consensus, promotional methods for the standard. The principal voluntary standards organizations of concern to fire safety are the twenty-five organizations referenced in the National Building Code.

3.6.4 General Acceptance of Standards in the United States

Unlike other developed nations, the U.S. standards system is highly fragmented. Mandatory building fire codes are enacted and enforced at the local level (town, county, city). The United States does not have a central code dealing with fire safety.

3.6.5 Enforcement and Incentives

There are two principal ways to impel people toward fire safety: through the police power of government at various levels and through economic incentives.

3.6.5.1 The Police Power

At the local level fire safety is achieved by the absolute power of building permits and by the licensing authority coupled with the power to inspect that regulates occupancy. The foregoing applies to buildings. Vehicles may be permitted or not permitted to be built or operated according to their degree of conformity to established fire safety standards. This applies to ships, aircraft, and land vehicles. The police power in this area can be exercised by state (notably California) but, more generally, it is exercised by the Federal Government which also controls such hazards as flammable fabrics and consumer goods generally.

3.6.5.2 Incentives

There are, however, economic pressures that are brought to bear to direct people toward fire safety. An important incentive is the size of the insurance premium that inversely reflects the degree of fire safety. On the other hand, as brought out in a recent study, "In America, fire department costs are typically funded from property taxes. This has resulted in a negative tax incentive situation which discourages the use of fire safe construction and installation of private fire protection equipment in buildings. Large buildings without on-site fire protection installations require public fire departments to focus an inordinate amount of resource to deal with the high fire flow requirement. This usually results in the small building owner subsidizing the fire department cost to structures other than his own." Other disincentives are increase in property assessments resulting in higher taxes when improvements that reduce the fire hazard are made to a property.

3.6.5.3 General Conclusions and Recommendations

The fire safety of our environment can be improved and losses from unwanted fires can be reduced significantly only as progress is made in identifying test methods which are practical, meaningful, and useful in the development of safer materials, systems, and designs.

The development of test methods to date has been primarily guided by ad hoc pragmatic considerations, because adequate understanding of the underlying processes of fire physics, dynamics, and chemistry has been lacking.

The development of most existing tests was based on observations of the burning behavior of natural polymeric materials, primarily cellulose. When these tests were applied to synthetic polymers, particularly to thermoplastic materials, results were misleading. For many tests, the difference in physical response between thermoplastic and non-thermoplastic materials precludes direct comparison of results and meaningful rankings.

Most existing laboratory test methods reflect the flammability of material and the relative hazards associated with it in a limited, relative, and often inaccurate manner when the data are extrapolated from the laboratory to the real world.

Characterization of the fire safety of a given material by any one laboratory test method is not a valid basis for material selection for design use.

The results of several judiciously selected laboratory test methods can, in some cases, provide useful indications of probably relative fire hazard and materials' ranking over a limited range of fire conditions.

Fire safety (or fire hazard) assessment involves determination of many aspects of the flammability behavior of a material, including smoke evolution, toxic gas emission, and fire endurance. Only an experiment designed to measure each pertinent parameter might characterize fire safety for a given material or system.

Tests exist for each of the above parameters. They vary widely with respect to age, sophistication, and merit. Tests for toxic gas emission are less developed than others.

System testing and laboratory simulation of real fire situations are desirable but exceedingly complex approaches to fire testing. Problems of scale, and the multitude of variables to be considered, preclude a single definitive choice of test configuration. In most instances, such approaches to testing are useful only as pragmatic expedients for specific situations.

Full-scale testing of systems is essential when critical decisions regarding fire safety of materials must be made because laboratory tests, even in combination, cannot adequately describe behavior in real fire situations.

3.6.6 Specific Conclusions and Recommendations

Conclusion: Recognizing that all contributors to the fire protection system should be stimulated to maximum activity in the pursuit of fire safety, incentives should be sought to further these activities. Recommendation: Search out economic incentives such as favorable insurance rates, favorable tax treatment, or other societal incentives that favor improved fire safety.

Conclusion: Only the United States among the developed nations has a widely fragmented system of fire safety codes and standards. This contributes to avoidable losses of life and property. Recommendation: Attack the problem of fragmentation of fire codes by judiciously selecting a pattern from the practices of other developed countries and adopting their practices as seems desirable in the United States.

Conclusion: Codes and standards are only of value to the extent that they are enforced. Recommendation: Enforcement of fire codes and mandatory standards should be uniform and rigorous.

Conclusion: Recognizing that private, quasi-public (insurance) and public parts of the fire protection system should be maximized for fire safety, incentives should be sought to stimulate activities that improve fire safety. Recommendation: Search out economic incentives such as favorable insurance rates, favorable tax treatment, etc., that would favor activities looking toward fire safety.

Conclusion: Only the United States among the developed nations has a widely fragmented fire code situation leading to confusion and avoidable losses of life and property. Recommendation: Attack the problem of fragmentation of fire codes by judiciously selecting a pattern from the practices of other developed countries.

Conclusion: Codes and standards are only of value to the extent that they are enforced. Recommendation: Enforcement of fire codes and mandatory standards should be uniform and rigorous.

CHAPTER 4

SMOKE AND TOXICITY

4.1 Introduction and Scope

4.1.1 Scope and Exclusions

This chapter will consider the health hazards arising as a result of the thermal decompositon and combustion of polymeric materials in fire situations. The current state of knowledge about the nature of the products involved and their potentially harmful effects on biological systems will be reviewed.

This report will not concern itself with the specific physiologic reactions, thereby, and prognosis associated with body surface burns or the health hazards of raw polymers.

4.1.2 Philosophy of Fire Toxicology

All chemicals are potentially capable of producing harmful effects on biological tissue. In any fire situation, it is possible to observe toxicologic effects dependent on the quantity of compounds involved, time of exposure, and the proximity of the exposed organisms to the fire.

4.1.2.1 Fundamental Concepts of Experimental Toxicology

Combustion toxicology is inherently complex because relatively minor variations in procedure may cause major changes in the actual dose of toxicants produced. Such procedural variations include the heat flux used, whether combustion is flaming or nonflaming, the duration of heating, physical configuration of fuel, relative mass of the material degraded, etc. To cite the most prevalent example, fuel burned in a hot fire with plenty of oxygen will very often yield mostly carbon dioxide, but the same fuel burned in limited oxygen may yield mostly carbon monoxide; the resultant toxicities are dramatically different.

In order that data can be compared at different times or by different laboratories, the methods of toxicologic testing should be comparable, and the product undergoing tests should be thoroughly identified.

Whenever toxicity differences are ascribed to different polymers, these differences should be of practical significance and not just statistical differences of little practical importance.

4.1.2.2 Life-Threatening Conditions Induced by Fire

The direct clinical consequences of fires stem from at least two primary sets of conditions – *elevated environmental temperatures and toxic combustion/pyrolysis* conditions. Elevated temperatures can create a stress ranging from trivial to lethal. "Toxic" combustion/pyrolysis conditions can be categorized as specific chemical combustion products, decreased oxygen concentration, and smoke.

4.1.2.3 Clinical Toxicology of Products of Combustion

The hazards to human health associated with exposure to fires involving polymers can be evaluated best when epidemiological data from humans are available to supplement animal data.

The current state of the art in regard to establishing a cause of death always involves an educated interpretation of the available pathologic and toxicologic evidence.

4.1.2.4 Fire Hazards

A real fire situation results in a continuously changing variety of toxicologic hazards to human health.

4.1.2.4.1 Hazards Common To All Fires

In a man-made environment, all fire situations yield heat, decreased atmospheric oxygen concentration, and carbon monoxide.

4.1.2.4.2 Hazards Related To Ventilation and Heat

As opposed to the situation in well-ventilated fires, fires that occur under poorly ventilated conditions tend to result in:

- *Significantly decreased* atmospheric oxygen.
- Limitation of the rate of combustion, thus a *limited production* of thermal energy.
- Increased carbon monoxide and other toxic or irritant products of *incomplete combustion.*

In any real fire situation, these three categories of factors probably never exist as separate entities. Their health hazard lies in their potential for additive or synergistic interaction.

4.1.2.4.3 Hazards Related To Specific Polymeric Substances

Polymers that undergo pyrolysis or combustion may yield varying amounts of heat or smoke, and/or additional specific toxicologic agents, such as hydrogen chloride, sulfur dioxide, hydrogen cyanide, and nitrogen oxides.

Nearly all polymers undergoing pyrolysis or combustion produce multiple products in *continuously varying* amounts. Not all of these combustion products have been toxicologically characterized. Consequently, comparison of the health safety aspects of these combustion products can be made *only* by conducting toxicologic tests with the actual polymers using experimental animals, so that the resulting data can be related to man.

4.2 Summary of Conclusions and Recommendations

4.2.1 Introduction

Since the first wood fires were lit, combustion and thermal degradation of polymeric materials has produced toxic products capable of exerting a variety of physiological effects. It is recognized that fire causes a health hazard for both accidental victims and fire fighters. Fire may be fatal. Surviving fire victims may carry the physical or emotional experiences of a single brief encounter with fire for years. However, exposure of fire fighters is chronic – less intense but steadily repeated.

4.2.2 General Conclusions and Recommendations

In any typical, uncontrolled fire, the toxic environment is continually changing. This environment is a function of the amount and ingredients of the polymers that are being burned and decomposed. This actually controls the toxicity of the fire. No condemnation can be made of synthetic polymers *per se.*

Present test methods are only partly adequate to evaluate polymeric materials. Also, toxic decomposition conditions generated by burning a single material may be greatly altered in the presence of other burning materials. It is erroneous to consider that simple toxicity tests offer a precise hazard evaluation. Rather, they offer a guide that may permit logical rating of toxicity under specified conditions and permit the recognition of unusually toxic pyrolysis or combustion products.

4.2.3 Specific Conclusions and Recommendations

4.2.3.1 Analysis of Combustion Atmospheres

Conclusion: Various methodologies for the experimental study of fire toxicity are currently in use. A standard, reproducible test is needed to generate orderly ranking of experimental data that can be correlated to fire scenarios. However, the state of the art is such that it is not now feasible to stipulate specific test methods in detail. Recommendation: Develop guidelines listing key factors for fire toxicity tests. These should serve as the basis for a future test protocol(s).

Conclusion: A given polymeric material in one use may not present the same hazard in another application. Recommendation: Evaluate the relative risk of a candidate product, based on final composition, end use, or quantity that can reasonably be anticipated to be present in a given fire situation.

Conclusion: The chemical nature of the product being burned may or may not be particularly critical to toxicity – carbon monoxide and carbon dioxide are universal but some unusual toxicants generated by polymer combustion may be highly or even extremely toxic. Recommendation: Perform bioassays. Use chemical analysis of pyrolysis/combustion products as an interpretive guide to toxicity phenomena, not an absolute rule *per se.*

4.2.3.2 Experimental Biological Effects

Conclusion: Animal and human data indicate that carbon monoxide is still the major problem in fires. However, other toxicants, particularly smoke particulate, may obviously be involved. Recommendation: Conduct further study to assess the nature and relative importance of these secondary toxicants. Evaluate smoke for obscurant properties as well as direct tissue irritation.

Conclusion: No agreement has been reached on suitable test end points. It is now recognized that mortality alone is inadequate. At present, the most promising incapacitation end points are sensory irritation and neuromuscular dysfunction. Recommendation: Expand evaluation of these end points.

4.2.3.3 Human Data

Conclusion: With the exception of carbon monoxide, data on the epidemiological aspects of fire are too meager to compare the contribution of various factors to the fire toxicity potential as a whole. Quantitative data are particularly scarce on subjects such as amount and nature of material in a "typical room" burn, time of exposure, effects of physical conditions and general health, acute versus chronic response, and delayed effects from fires. Recommendation: Make detailed systematic studies directed to gathering these types of data from files and fire victims.

Conclusion: Increasingly, prevention of exposure to a fire environment, where feasible, is seen as the key to preventing injury. Recommendation: Refine and popularize early warning systems until their use is the rule rather than the exception throughout confined areas. Design highrise and densely populated buildings with a "fire fighting plan" emphasizing fire containment by structural design and devices such as automatic sprinklers.

4.3 State of the Art

4.3.1 Introduction

In the quarter-century since a basis for the toxicology of fire was first expressed, formal study of the life hazard from fire has progressed in two broad directions:

- Surveys of fire injuries and deaths.
- Increasingly sophisticated testing, both laboratory and large-scale.

This chapter contains a review of basic clinical and experimental data, analytical methods pertinent to current concepts of combustion toxicology, and an overview of experimental design.

4.3.2 Causes of Death and Incapacitation in Fires

Scientific study of fire fatalities began as an aftermath of the 1929 Cleveland Clinic fire. Since that time mass fatality fires and wars have enabled investigators to identify some specific causes of death and incapacitation. These factors can be categorized as *thermal, chemical,* or *extrinsic.* At any given moment in a real fire the net sum of these varying factors represents the life hazard.

4.3.2.1 Hyperthermia

Hyperthermia (abnormally high body temperature) is frequently over-shadowed by other more apparent injuries and should be suspected in any case where body surface burns are more than 30%. Tests with large animals having a surface area-mass relationship near that of humans show relationship between exposure time and temperature is hyperbolic. If exposure time is increased, a lower source temperature produces a given threshold of injury.

4.3.2.2 Toxic Decomposition Gases and Conditions

In any fire situation with mixed fuel, toxic combustion products are a heterogeneous mixture of gases, liquid droplets, and smoke particulate. Combustion may also cause oxygen deficiency, which exerts a toxic stress; hypoxia is, therefore, considered a toxic decomposition condition.

4.3.2.2.1 Conditions Producing Hypoxia

In fire situations carbon monoxide (CO), hydrogen cyanide (HCN), and deficiency of oxygen (O_2) are the three most common causes of hypoxia. Earlier literature regarding synergistic effects of chemicals is unclear, but a recent study describes a definite additive effect from CO plus HCN.

4.3.2.2.1.1 Carbon Monoxide

Carbon monoxide-induced hypoxia has been considered for years the major single cause of fire fatalities, but specific validating data were lacking until recently. Carbon monoxide (CO) is absorbed via the lungs into the blood where because of hemoglobin's greater affinity for CO than O_2 the amount of hemoglobin (Hb) available for oxygen transport is reduced.

4.3.2.2.1.2 Hydrogen Cyanide

Fatalities in aircraft fires and other enclosures with limited or no egress account primarily for the interest in HCN. Attribution of incapacitation and death to HCN appears very questionable in cases where egress has been prevented by mechanical means and where autopsy showed pulmonary irritation, smoke particulate in respiratory passages, and high COHb levels.

4.3.2.2.1.3 Oxygen Deficiency

Lowered atmospheric O_2 concentration is an obvious contributory factor to hypoxia. Symptoms of deficiency are insidious and vary with degree of deficiency, duration, and physical condition of subject. Generally, 10–15% oxygen appears to be critical atmospheric concentration below which serious impairment can be expected.

4.3.2.2.2 Irritant Gases and Vapor

In fire situations, gases and vapor affecting the eyes and respiratory tract may cause tissue injury or make it difficult to see or breathe. This may result in a delayed exit until victims are overcome by other injuries such as CO poisoning or burns.

4.3.2.2.2.1 Sensory Irritants

These chemicals inhibit respiration and cause one or more typical reflex actions associated with stimulation of trigeminal nerve endings. Such reflex actions include: burning sensation in the nose or eyes, moist facial skin, sneezing, coughing, and tears (lacrimation). Typical sensory irritants are hydrogen chloride, sulfur dioxide, acrolein, and ammonia.

In fire situations, one sensory irritant that has lately received increasing attention is hydrogen chloride (HCl), usually attributed to thermal decomposition of polyvinyl chloride (PVC) whose main decomposition products are HCl, CO, CO_2, and benzene.

4.3.2.2.2.2 Pulmonary Irritants of the Lower Respiratory Tract

These irritants typically increase respiratory rate as a result of "stimulation or sensitization of various nerve endings of the lower respiratory tract," and such action is usually without immediate painful sensation. "True pulmonary irritants" are ozone, nitrogen dioxide and phosgene, since the chemicals exert their primary action solely on the lower respiratory tract.

4.3.2.2.2.3 Bronchoconstrictors and General Respiratory Irritants

Bronchoconstrictors are chemicals that induce increased resistance to airflow within the conducting pathways of the lung, either by nerve stimulation on the smooth muscle airways or by liberation of histamine. (This constriction may be painful, but pain may also develop from mucosal swelling, a separate process.) Typical examples are sulfur dioxide, ammonia, and sensitizing allergens such as foreign proteins, toluene diisocyanate, or histamine.

The term *respiratoy irritant* indicates that an inhaled chemical can act as sensory irritant, pulmonary irritant, and bronchoconstrictor with little difference between the effective concentration for sensory irritation and pulmonary irritation. Typical examples are chlorine and ketene.

Sensory irritants characteristically decrease respiratory rate through action on the upper respiratory tract, and pulmonary irritants increase respiratory rate through action on the lower respiratory tract.

4.3.2.2.3 Smoke Particulate

In fires using mixed fuels or natural polymers, smoke particulate is a tacky, oily aggregate of carbon particles with adsorbed combustion products that are usually irritants.

Smoke particles may be small enough to be inhaled deeply into lungs. These particles can convey adsorbed irritants or residual heat into any part of the respiratory tract that they contact.

Smoke may also cause obscuration of vision that can result in a life hazard if exits cannot be located.

4.3.2.2.4 Miscellaneous Gases and Vapors

Certain gases and vapors may be adsorbed on smoke particles or be airborne in the fire effluent. Carbon dioxide from complete oxidation of carbon atoms might be the most common product in fire situations. Other gases with anesthetic or drug-like properties, such as propane or benzene, may be evolved in relatively small amounts as pyrolysis products.

4.3.2.3 Additive and Extrinsic Stress Factors

Factors in this category are not an inherent part of fire toxicity but yet may lead to severe injury or death, either directly or by aggravation of other injuries.

4.3.2.3.1 Panic and Emotional Stress

Statistics are almost nonexistent. Panic, fear, or other emotional strain can result either in shock requiring direct medical attention or lead to incapacitating behavior.

4.3.2.3.2 Trauma

Trauma may be caused by falling objects, jumping to safety, or panic. Trauma is most likely in confined spaces such as buildings or aircraft with limited exit facilities.

4.3.3 Fire Fighters

A 1976 report analyzing the deaths of 101 fire fighters "in-the-line-of-duty" found that 45 died from "heart attack." The average fire fighter victim dying from heart disease was 51 years old with 22 years of service. Exposure to smoke and toxic fume inhalation, stress, and overexertion were considered the main contributory causes of these "heart attack" deaths.

4.3.4 Experimentation in Fire Toxicology

Modern study dates from 1940 when it was shown that thermal damage to the respiratory tract and chemical inhalation injury often caused fire deaths without body surface burns. An extensive series of studies that explored the primary variables of hyperthermia, CO, anoxia, and CO_2, as well as the synergistic interaction of these variables and the superimposed effect of small amounts of hydrogen cyanide (HCN), nitrogen dioxide (NO_2), and sulfur dioxide (SO_2) indicated that hypoxia and heat increased the toxicity of CO while CO_2 did not appear to increase CO toxicity. Very small amounts of HCN, NO_2, and SO_2 produced "a considerable increase in toxicity with respect to hypoxia, CO and CO_2."

Analytical techniques vary widely from small-scale laboratory tests to room tests. The basic approaches that have been used in chemical analyses are: a) detailed analysis of decomposition products using gas chromatography (GC) or mass spectrometry (MS) and b) analysis for specific selected products.

Both physical and chemical characteristics of the airborne particulate influence the life hazard in fires. Depending on size, adhesiveness, and composition, smoke particles may obscure vision, physically coat the respiratory tract, and act as carrier for toxic gases.

At present, no direct correlation between biologic, chemical and physical characteristics of any type of fire is known. Data comparing smoke density and CO indicate that the relationship is not simple.

4.3.5 Analytical Test Methodology

Sampling and analysis of thermal decomposition products are formidable tasks. In a few special cases – determination of hydrogen cyanide and hydrogen chloride in air – development has been initiated. In most instances, sampling and analytical methods previously developed for studies of air pollution, automotive emission, waste water, and pharmacology could be employed.

4.3.5.1 Analysis of Nonabsorbing Gases and Vapors

Carbon dioxide, carbon monoxide, oxygen, and most hydrocarbons are the easiest products to sample because they are less prone to interact with particulate and water. Because of their rapid generation rate in structural fires and the toxicological significance of carbon monoxide, continuous analysis methods are preferred. Optical systems such as nondispersive infrared (NDIR) can be used for the organic analyses. As an alternative, GC can be used, but continuous monitoring is not possible by this method.

Detailed analysis of the several hundred organic compounds that may be pyrolyzed requires high resolution GC confirmed by MS analysis of the trapped or separated chromatographic effluents.

4.3.5.2 The Troublesome Acid Gases

Hydrogen chloride, hydrogen fluoride, hydrogen cyanide, sulfur dioxide, and nitrogen oxides can all be absorbed by particulate, tars, and condensed water of combustion. In most instances, continuous monitoring methods are not available. Moreover, no general approach such as ion specific electrodes can be used for these gases in all instances, due to interferences and lack of sensitivity.

4.3.5.2.1 Hydrogen Chloride

Hydrogen chloride is one of the principal products from pyrolyzed or combusted polyvinyl chloride (PVC). Most investigators sample for hydrogen chloride by scrubbing the smoke through a liquid impinger containing dilute caustic (0.1 N sodium hydroxide). The impinger sample is then analyzed for chloride using an ion specific electrode or spectrophotometric technique.

4.3.5.2.2 Hydrogen Cyanide

Hydrogen cyanide may be produced from any nitrogenous polymer, Also, trace amounts are reported in the combustion of cotton and paper and appear to be formed by fixation of atmospheric nitrogen. This gas is highly lethal and causes impairment of motor function as a result of "cytotoxic anoxia" in very low concentrations in the blood (1/6 the lethal blood level).

Sampling and analyzing HCN is difficult since this gas is absorbed or absorbed by the moisture and smoke particles in a fire atmosphere as well as tubes and equipment. Interference is another problem. For critical work, many workers prefer impinger sampling for collection and analyses of cyanide.

4.3.5.2.3 Hydrogen Fluoride

Hydrogen fluoride may be formed as a combustion or pyrolysis product of fluorocarbon polymers. Analytical sampling systems similar to that for hydrogen chloride may be used. A fluoride ion electrode can be used to monitor impinger samples.

4.3.5.2.4 Oxides of Sulfur and Nitrogen

Several investigators have reported sulfur dioxide (SO_2) to be formed from wool, and nitrogen oxides (NO), as nitric oxide (NO) and nitrogen dioxide (NO_2), from a wide variety of polymers. Procedures based on the chemiluminescent reaction can be used for SO_2. For NO, NDIR as continuous monitoring and Saltzman's reagent as a batchwise method are available.

4.3.5.3 Smoke Obscuration

All fires produce smoke, which may be thick or thin, light or dark. Any appreciable quantity invariably is irritating and obscures light and vision. No universal or even general standard exists at present for measuring smoke obscuration, however, in this country, preference has been shown for the National Bureau of Standards (NBS) smoke chamber.

As would be expected, flame retardants increase carbon in the pyrolysis residue and decrease the yield of volatile aromatic products.

4.3.5.4 Biological Analyses

The easiest endpoint for biological studies of smoke is "counting dead rats or mice" (cessation of respiration). Monitoring blood levels of the asphyxiants carbon monoxide and hydrogen cyanide shows whether these agents caused, or signficantly contributed to, clinical distress or death. In practice "biological" tests frequently use a coordinated approach, combining biological and analytical methods.

4.3.5.4.1 Sensory Irritation

Reflex-induced modifications of the respiratory pattern provide a sensitive means to detect potential irritants and identify concentrations at which they will be irritating.

Thermal decomposition products of complex polymers may contain a variety of irritants, which can be identified by chemical means. Pulmonary irritants may not immediately cause obvious physiological changes at mild to moderate levels.

4.3.5.4.2 Hypoxia

Neuromuscular dysfunction caused by CO and HCN has been studied in rats by failure in reflex tests and failure in trained activity tasks using an activity wheel or rotorod. Investigations of the degrees of intoxication that are compatible with survival include three levels, a) behavioral functioning needed for escape, b) hypoxia or shock requiring immediate first-aid to avoid death; and c) morbidity following acute intoxication.

4.3.6 Experimental Design in the 1970's

Contemporary experimental design for combustion toxicology involves five major concepts: fire statistics, chemical and physical analyses, biological response, a combination of chemical/biological testing, and comparative material studies. Use of the various concepts depends on test goals.

4.3.6.1 Fire Statistics

Fire statistics provide the direction and impetus for studies with animal models and may also validate conclusions. For example, CO was long suspected as the main cause of fire deaths, and human studies have now demonstrated this to be so.

4.3.6.1.1 Forensic Pathology and Toxicology

When toxicologic analytical investigation is performed, only one combustion, CO, has been routinely measured in the blood of fire casualties. If the blood is more than 50% saturated with CO even though the body is severely burned, it is highly probable that the person died of CO before physical burns occurred. Recently it has become increasingly clear that this gas, although ubiquitous in fire situations and not to be underrated as a cause of death, may represent only a part of the total toxicity of combustion mixtures. Consequently, some forensic laboratories now estimate blood cyanide, arising from the combustion of nitrogen-containing materials, in addition to carboxy-hemoglobin saturation.

One fundamental proviso underlies the use and usefulness of all such toxicological data — measurements must be accurate; they are not easy to perform. Volume 3 should be read for a discussion of this point.

4.3.6.1.2 Clinical Investigations

In this country and abroad, a number of groups have established reputations for their studies of humans who have escaped or have been rescued from fires.

Useful toxicological information can be derived from observations on firefighters or fire-survivors who have suffered severe chemical exposure.

4.3.6.2 Chemical Analysis of Decomposition Products

Analyses of fire-generated toxicants have generally relied either on detailed chemical assessment of the range of thermal decomposition products or measurements of selected products. The first is highly complex due to the wide range of physically different products (particulate, mist, and gas) requiring quite different methods of analysis.

The alternate approach with selected products is far simpler but in a number of cases will be based on questionable assumptions. For reasons of cost, measurement of selected products is sometimes the only feasible approach but the value of the data is less.

It is highly improbable that a new species of highly toxic agent will be found except in those cases whereby new forms of fire retardants or dyes may be unknowingly involved in the test. In that case the experienced toxicologist would be alerted because of the inconsistent nature of the observed effects on the test animals. Therefore, it may be an unnecessary and even undesirable procedure to involve complicated chemical monitoring equipment in routine animal tests. Even with the use of GS/MS equipment there is rather minimal chance that small amounts of new, highly toxic substances would be revealed. The most rational and economic procedure would be to reserve the elaborate techniques such as GS/MS monitoring for use when the toxicologic data demonstrates its need.

4.3.6.3 Tests for Inhalation Toxicity (Biological)

Tests with experimental animals provide measurements of biological response, which fall into three main types:

- Mortality with or without accompanying clinical signs.
- Incapacitation, usually either as loss or activity or reflex response.
- Sensory irritation

Mortality is dramatic and relatively easy to measure, but it is increasingly considered inadequate as the sole criterion in fire situations.

In instrumented animals neurological, physiological or biochemical changes that indicate impending incapacitation or loss of purposeful activity could be selected. Loss of coordinated neuromuscular activity would effectively end an animal's efforts to escape from the hostile.

Sensory irritation is, in a sense, a measure of incapacitation also. More precisely, it involves a measurement of concentration and time when reflex changes leading to incapacitation begin, not when animals are totally incapacitated. Inherently, appraisal by sensory irritation provides a measure of warning that is lacking and has to be estimated when mortality and/or loss of useful function are chosen as criteria.

4.3.6.4 Combined Analytical and Biological Testing

Most of the more recent combustion toxicology experiments conducted with animal exposures have used some chemical analyses, but the term "combined" has been suggested to indicate a distinct dual approach. Such determinations, however, may involve a number of problems, especially when particulates are present.

4.3.6.5 Comparative Material Studies

Comparative material studies of natural and synthetic polymers is still the basic approach to evaluation of candidate synthetic polymers. But this concept must be placed in perspective. It is unrealistic to demand, for example, that all proposed building materials be "no more toxic than wood." Wood varies in composition – even samples of specific type, such as Douglas fir or white pine, are by no means chemically identical. It is extremely difficult to define what constitutes suitable combustion/exposure conditions. Experimental data now available dictate caution in any proposed standardization. Several major pitfalls are identified below.

4.3.6.5.1 Method of Thermal Degradation

Materials can be thermally degraded in three principal ways: anaerobic pyrolysis, oxidative pyrolysis ("smoldering"), and flaming combustion. All conditions can and probably do occur at some point, at some time, in a real fire, and change in prominence as the fire proceeds.

Test results indicate variation in "best-to-worst" toxicity ranking depending on whether a static chamber "pyrolysis" method or a dynamic chamber combustion method was used. Rankings obtained by the two different methods indicate that no one method provides the most severe test for a candidate synthetic polymer.

4.3.6.5.2 Temperature Range

Most materials, natural or synthetic, yield volatile, potentially toxic fragments at around or even below 400°C.

For experimental screening, however, temperatures as low as 400°C to 500°C may be inappropriate because many polymers will not sustain their own spontaneous pyrolysis in this region, and the most resistant will only partially decompose.

A temperature of 850°C may present difficulties and introduce the need for separate burning and longer transfer pathways to protect the animals from heat.

A temperature of about 600°C constitutes a severe test for organic materials, as many polymers decompose rapidly at this temperature, however, variations around this temperature may lead to significant differences.

4.3.6.5.3 Time Frame

The total elapsed time for burning and exposure should take into consideration practical aspects – the rapidly developing fire. If time to incapacitation and time to death are to be observed, the sample size (concentration) should be so adjusted that one or both occurs within 30 minutes.

The inherent resistance of a polymer to thermal decomposition will influence the rate of burning under the conditions selected, and this is an uncontrollable variable in the elapsed time to an observable toxicological reaction.

4.3.7 Conclusions and Recommendations

Conclusion: Carbon monoxide is the major and most common chemical responsible for toxicity from the pyrolysis/combustion of polymers. Some studies involving products from specific polymers indicate a discrepancy between the quantitites of CO found and mortality. Recommendations: Wherever practicable, in experimental studies monitor levels of CO in both the atmosphere and the blood of animals. Obtain in clinical studies additional data regarding the role of CO as the involved toxicologic agent. In both experimental and clinical studies, routinely evaluate the role of other possible factors.

Conclusion: Relatively little is known about either combustion products other than CO that may develop or their total biologic effect (synergy). Still less is known about the time in the fire sequence when such decomposition products may reach biologically significant levels. Recommendation: Consider in search programs for studying combined effects of combustion/pyrolysis products, where feasible, whether relatively uncommon decomposition products develop before or after the more frequent problems of CO, O_2 deficiency, and excessive heat.

Conclusion: Due to the variety of methods in use in the experimental evaluation of the toxicity of pyrolysis/combustion products, as well as the failure of some of the investigators to conform to basic toxicologic principles in the design and conduct of their studies, comparison between studies is frequently impossible. Recommendation: Cause all protocols for studies and reports of data to conform to acceptable toxicologic practice, including the use of a sufficient number of animals to give statistically valid results. Make routine screening protocols uniform to the maximum practicable extent so that comparisons in test results can be made.

Conclusion: The past experimental toxicologic studies on animals have involved either measurement of irritant action on the airway, measurement of neuromuscular incapacitation, or measurement of lethality. At the present time, preference for any one type of measurement as the most significant for extrapolation of toxicity to man lacks adequate validation. Recommendation: Perform studies to evaluate the relationship between the various feasible measurements to resolve the question of which measurement is most appropriate.

Conclusion: Clinical investigation of the toxicity aspects of fire casualties has been conducted in only a few centers. Available data indicate that chemical (combustion product) toxicity is responsible for substantially more incapacitation than is the physical burn phenomenon. Little or no information exists about the role of smoke obscuration in escape. Recommendation: Expand effort in the areas of clinical epidemiologic and toxicologic studies of fire casualties. Fund centers having capability in these study areas.

Conclusion: It is not possible at this time to brand a product as "good" or "bad" on the basis of isolated toxicologic tests for evaluation of polymers and their combustion products. It is feasible to provide a spectrum that reveals relative toxic potential among materials designed for the same end use. Recommendation: Develop this spectrum and periodically update it by an appropriate, technically qualified group. Give manufacturers, users, and regulatory agencies an opportunity to make inputs to this group.

4.4 Gaps in Knowledge of Smoke and Toxicity

4.4.1 Introduction

The gaps in our knowledge of the health and safety effects of smoke and toxic products from the combustion of high polymers are so numerous and ill-defined that they can only be discussed in broad categories.

4.4.2 Smoke and Acute Inhalation Toxicity

Smoke may cause two kinds of hazard for persons involved in a fire. First, it reduces visibility sometimes to the point where victims cannot see exits and may be unable to escape from the fire. Second, smoke particles are irritating to the eyes and respiratory tract, probably due mostly to gases adsorbed on their surfaces.

4.4.3 Toxicity Testing Parameters

At the present time no generally accepted procedures for fire and smoke toxicity screening of polymers are available. Most of the component hardware and subprocedures are available from various fire toxicity research laboratories.

Thirty minutes is probably an adequate exposure time since relatively few real fire situations last more than 30 minutes without intolerable heat becoming the limiting factor for human survival. Animals should be observed for a minimum of 14 days after exposure to detect any residual injury in the survivors.

4.4.4 Role of Chemical Analysis

Elaborate analytical procedures are not recommended for screening tests. However, analyses for such "standard" gases as oxygen (O_2), carbon dioxide (CO_2), carbon monoxide (CO), and anticipated incapacitating gases, for example, hydrogen cyanide (HCN) are desirable.

4.4.5 Multiple Simultaneous Exposures

Toxicology studies generally are, and have been, almost totally concerned with single chemical compounds. It has long been recognized in real life situations, and especially in the fire scenario, that these kinds of exposures are complex with exposure to multiple toxic materials occurring simultaneously.

Components of the combustion products can interact with each other to produce a new mixture having equal, less, or more toxicity than before their interaction. These interactions may be of a chemical, physical, or biological nature.

Much research of a chemical and biochemical nature is needed in order to identify combustion products, possibilities for interactions, and overall toxic potential in the dynamic fire exposure situation.

Combinations of gases must be more thoroughly investigated if a rational hazard index is to be derived for a polymeric material involved in fire.

4.4.6 Retardants and Extinguishants

A comparatively small number of additional materials may be intentionally associated with a fire situation. Some, the fire retardants, are added to the potential fuel in advance in order to reduce its flammability or retard its rate of combustion. Others, the extinguishants, are introduced after the fire has started in order to put it out.

Retardants can be considered as another class of materials added to fuel resulting in problems from interaction as well as their intrinsic toxic and physiologic effects.

Extinguishants present a dual toxicity problem, that of the materials themselves and that of their decomposition products singly and in combination. The toxicity of extinguishants can vary widely.

4.4.7 Interpretation of Results

The purpose of considering the toxic effects of combustion products of high polymers is to make comparative judgments of their potential ability to cause death, incapacitate leading to death, or permanently injure survivors.

It is therefore desirable that a fire toxicology test include the added stress of heat and reduced O_2 concentration in its basic design in addition to the rates of evolution of combustion products from various polymeric systems.

The total effect of combustion products upon the body, both in magnitude and speed of occurrence, should include the rates of various processes within the body and the body's reserve capacities. We know very little about any of these dynamics of the toxicology and physiology of the components of the combustion products or of the total mixture.

The behavior of persons involved in a fire situation has often been responsible for injury. During fire toxicology tests, test animals should be carefully observed for behavioral changes and any evidence of incapacitation as well as morbidity and mortality.

4.4.8 Coordination and Information Processing

There is an immediate need for reliable comparative information on the fire aspect of polymers to be used in the wise selection of materials for construction. The methods or organizations involved in fire tests need to be standardized so that results can be correlated for greater understanding of the toxic hazards involved.

However, a comprehensive, rapidly responsive, widely disseminated system of information exchange of past data, recent results and planned research or testing would avoid unnecessary duplication of effort, assure comparability of data, and help identify issues of the highest priority.

4.4.9 Conclusions and Recommendations

Conclusion: More research is needed on the dynamics of the physiologic and toxic effects of polymer combustion products. Recommendation: Develop a research program in these specific areas: (a) combined effects of irritant gases and other toxic materials produced simultaneously in fires; (b) entrapment, distribution, and excretion of particles inhaled rapidly through the mouth, as occurs in fire situations; (c) theory and methodology for separating experimentally and recombining predictively the fire stresses of toxicity, heat, and hypoxia.

Conclusion: Tests for measuring the acute toxic effects produced by polymers in fires should be comparable among different laboratories on the basis of specific chemical and biologic indices. Recommendation: Develop standard toxicity test procedures for burning conditions and exposure conditions. Include in these procedures chemical analyses conducted simultaneously with the animal experiments, plus measurements of the loss of capacity to escape and other behavioral effects.

Conclusion: Guidelines on the selection of materials are necessary to minimize the fire hazard in potential fire situations. These guidelines should include materials proposed for plasticizers, fillers, and fire retardants as well as the base polymer. Recommendation: Develop guidelines for selection of materials that consider: (a) reactions likely to occur during combustion; (b) combustion products likely to develop in potential fire situations; (c) likely rate of production of smoke and gases from polymeric materials as related to probable volume of end use of these polymeric materials in any given potential fire situation. (For example, polymers proposed as wall insulation might be a high volume use and polymers proposed as certain specialized mechanical parts might be a low volume or non-residential use.)

Conclusion: Obscurant properties of smoke have received relatively little attention. Recommendation: Explore any possible means of alleviating smoke obscuration. Such means should include improved selection of candidate materials and also techniques for use in potential fire situations. Techniques for dealing with obscurant effects of smoke should focus both on improved communication and physical dissipation of smoke through aggregation or collection of smoke particles.

Conclusion: Collecting and evaluating information on the many varied aspects of the life hazard in fire situations is difficult and is becoming increasingly more so as the complexity of combustion toxicology increases. Recommendation: Develop a modern, integrated fire safety information exchange.

CHAPTER 5

FIRE DYNAMICS

5.1 Introduction and Scope

This chapter specifically examines two interrelated approaches to improving fire safety: fire *dynamics* and fire *scenarios.* Like the other chapters, this is only a summary. Volume 4 of the Report should be consulted for a fuller exposition of the subject matter.

5.1.1 Fire Dynamics

Fire dynamics is the scientific description of fire phenomena in quantitative terms. Phenomena of interest include: ignition, flame spread, fire growth, maximum burning intensity, products of combustion, movement of combustion products through a building or other structure, detection, extinguishment, and effects of fire on humans.

5.1.2 Fire Scenarios

Fire scenarios are generalized descriptions of actual or hypothetical but credible, fire incidents. Frequently, a scenario is based on one (or more) actual fire incidents; however, since in most real fires complete knowledge of all pertinent events is impossible to obtain, scenarios are generally based on a combination of fact and speculation. Where the goal is to improve fire safety, an approach should be selected from a set of alternatives such as: replace existing materials with greater fire safety, but perhaps less suitable or more expensive characteristics; modify the existing design to reduce the hazard; increase the awareness of the users to hazards through education; control ignition sources, etc. To choose the approach wisely, an intimate knowledge of fire and fire behavior is required. An understanding of fire dynamics, coupled with the use of appropriate fire scenarios, can provide such knowledge.

To analyze each case, one clearly needs to know the chemical nature of the combustibles, the geometry of the fire compartment and adjacent compartments, ventilation factors, the location of occupants, the time lag until combustion products reach a human or automatic detector, as well as the possibilities of extinguishment, escape or rescue. Such cases require detailed scenarios.

The scenario approach helps clarify the nature and extent of hazards associated with products. In fact, a set of scenarios concerning a given product may permit generalizations on the safety of the product than can otherwise be formulated. Moreover, scenarios are needed to establish the relevance of test methods. For example, if an expensive full-scale fire test were to be performed on a product, considerable guidance in planning the test could be obtained by prior scenario analysis.

In scope, this chapter covers: the state of the art of fire dynamics and recommendations for its advancement; the need for the fire scenario approach to improve fire safety; detailed generalized guidelines for developing and analyzing fire scenarios; some examples of fully developed and analyzed scenarios; and many other examples of scenario outlines. Fire dynamic and scenario factors involving human response to fires are only briefly covered in this volume.

5.2 Fire Dynamics

5.2.1 Introduction

This section reviews the current status of fire dynamics as a scientific discipline, indicates potential practical applications, and recommends future efforts for the further development of fire dynamics.

If a material were expected to be exposed to fire risk in one and only one precisely defined set of circumstances (size, orientation, type of ignition source, method of applying ignition source, ventilation conditions, environmental conditions, etc.), it would be obvious how to test it for fire hazard.

Frequently, this idealized approach is not directly applicable because the material of interest may be exposed to fire risk in a wide variety of ways rather than in a simple well-specified set of circumstances.

As a result, many fire test methods are considered to be highly unreliable indicators of hazard. For many other cases, no appropriate test methods exist. Tests are particularly inadequate in regard to fire spread and growth, as well as noxious gas and smoke production.

5.2.2 The Scope of Fire Dynamics

What is fire dynamics? In broad terms, it involves quantitative descriptions of fire phenomena on a scientific basis. An understanding of these phenomena may be obtained on any one of four levels (see Section 3.5.2).

On the most basic level, we consider fire to be governed by a combination of effects tracing back to certain established fields of scientific study such as: the chemical thermodynamics and stoichiometry of combustion; the chemical kinetics of pyrolysis and combustion reactions; the transfer of combustion energy by conduction, convection, and radiation; as well as the motion of combustion gases as affected by buoyancy, thermal expansion or mechanical force.

On the second level, we may break fire down into a series of phases or stages, and consider fire dynamics to be an analysis of any of these phases:

Smoldering
Spontaneous ignition
Piloted ignition
Horizontal or downward flame spread over solids
Upward flame spread over solids

Flame spread over a liquid below its flash point
Flame spread over a liquid above its flash point
Burning rate of liquid pool
Burning rate of a solid slab
Formation of toxic species in a diffusion flame
Formation of aerosols in a diffusion flame
Radiation emitted by a diffusion flame
Extinguishment by heat loss
Extinguishment by reduction of oxygen
Extinguishment by chemical inhibitors

On the third level, we consider fire dynamics to be concerned with complex processes generally involving interactions of two or more phases given in the foregoing list. Some examples of such fire dynamics are as follows:

- Burning rate of an object as influenced by radiative feedback from the environment
- Burning rate of an object in a non-combustible compartment as influenced by ventilation of the compartment
- Generation of incomplete combustion products as influenced by either of the above burning rate conditions
- Mutual interactions of two adjacent burning objects
- Spontaneous ignition of pyrolysis gases from a hot object as influenced by turbulent free convection and mixing
- Properties of smoke from a fire in a compartment as influenced by the mixing, cooling, and "aging" or agglomeration, which occurs in the interval between smoke generation and its arrival at a detector station
- Adsorption of toxic gases from fires by aerosols from the same fires, as cooling occurs
- Analysis of radiant emissions, transmission, and adsorptions in a compartment at a pre-flashover stage of fire
- The effect of physical scale on fire turbulence, on fire radiation, and ultimately in fire behavior
- Delineation of relative effects of chemical kinetics and physical factors on a fire near an extinguishment condition
- Determination of the source of undesirable products of incomplete combustion in a fire; i.e., surviving initial pyrolysis products vs. products formed in gas-phase reactions in or near a flame
- Effects of long-term exposure of materials to various ambient conditions on subsequent fire behavior
- Identification of the mechanisms involved in the interaction of water spray with a fire.

Obviously, successful fire dynamics studies along these lines would strengthen the engineering judgment needed to validate test methods for the realistic determination of fire hazards.

The fourth level of fire dynamics involves interactions of fire phenomena with human response. Problems such as the following exist:

- Sensory detectability of fire (including smell and sound)
- Vision as affected by smoke and/or lachrymatory gases
- Panic or confused thinking as induced by fire phenomena
- Burn damage of skin by garments, particularly as influenced by the rate of flame spread, melting-dripping, etc.
- Toxicity including combined (synergistic) or sequential effects of various toxic species
- Toxicity, including prefire condition of victim (blood-alcohol content, heart or circulatory disease, etc.)
- Human ability to control fire at various stages of development as governed by training, equipment available, panic, etc.
- Actions of humans which lead to ignition.

While comprehensive, the foregoing four levels of approach to fire dynamics exclude phenomena related to the post-flashover development of a fire. They do not consider the spread of a fire from one part of a building to another; the responses of building components (ventilation system, windows, elevators, etc.) to fire; the spread of fire from one structure to an adjacent structure; and technology concerned with detection, communication and escape procedures.

5.2.3 Current State of the Art of Critical Fire Dynamic Elements

5.2.3.1 Introduction

In recent years, most of the current research on fire dynamics in the United States has been funded to various grantees by the RANN program of the National Science Foundation. Starting in fiscal year 1975, however, the newly created National Fire Prevention and Control Administration (NFPCA) took over the funding of these grants. From FY 1977, the Center for Fire Research of the National Bureau of Standards, has taken over support for an internal program, grantees, and contractors. Factory Mutual Research Corporation supports its own internal program of fire dynamics research. Several other government agencies support some fire research which includes, to a minor degree, fire dynamics studies. Table 1, based on estimates, is a representation of the total funding for fire dynamics research (In Table 1, research on the toxicology of fire gases is not considered to constitute fire dynamics).

Table 1 shows that the total available fire dynamics funding increased from \$2.39 million in FY 1973 to \$2.73 million in FY 1976, but, as of February 1976, is scheduled to *decrease* to \$2.03 million in FY 1977.

The *most critical* fire dynamics elements are *ignition* and *flame spread;* they will be briefly discussed here.

The ignition of a combustible material is the first step in any fire scenario and, therefore, is important to fire prevention. Furthermore, once a fire starts, the

Table 1. Fire Dynamics (FD) Research as Budget in U.S. Programs (Millions of Dollars)

FISCAL YEAR	NSF-RANN FIRE PROGRAM	NFPCA RES. PROGRAM	NSF + NFPCA			NBS CFR			FMRC INTERNAL PROGRAM			MISC. FD*	TOTAL FIRE DYN. RES.
			TOTAL	%FD*	TOTAL FD	TOTAL	%FD*	TOTAL FD	TOTAL	%FD	TOTAL FD		
73	2.00	0	2.00	60%	1.20	1.9	10%	0.19	1.0	50%	0.50	0.50	2.39
74	1.70	0	1.70	60%	1.02	3.1	10%	0.31	1.0	50%	0.50	0.50	2.33
75	0.65	1.12	1.77	60%	1.06	3.1	10%	0.31	1.0	50%	0.50	0.50	2.42
76	1.16	1.20	2.36	60%	1.42	3.1	10%	0.31	1.0	50%	0.50	0.50	2.73
77	0	1.50	1.50	20%	0.30	4.5	15%	0.72	1.0	50%	0.50	0.50	2.02

*Estimated

ignition delay times of other materials, coupled with flame spread, will affect the rate at which the fire spreads and develops. Hence the ignition delay time of these other materials is also important to fire development.

Most polymeric materials ignite and burn if they are sufficiently heated in the presence of air or other atmosphere containing sufficient oxygen. Consequently, from the viewpoint of fire safety, it is desirable to know how long it takes a particular polymer to ignite under various fire conditions, e.g., oxygen concentration, environmental gas temperature, heating rate, etc.

Historically, modern ignition theories have evolved from research directed towards determining the ignition mechanisms of composite solid propellents.

In the physical model illustrated by Figure 1, the process of ignition begins when the solid fuel surface starts to heat up as a result of heat transfer from the adjacent surroundings. As the solid fuel surface temperature increases in response to this heat input, fuel vapors are omitted from the surface. The fuel vapors mix by convection and diffusion with oxygen in the adjacent boundary layer, ultimately leading to heat liberation by chemical reaction in the gas if the gas temperature is high enough.

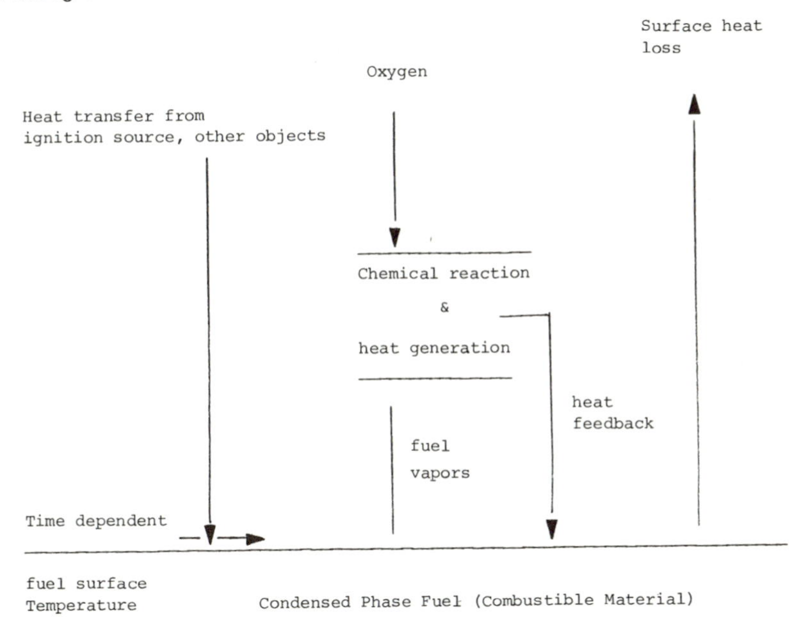

Figure 1. Physical process in gas-phase ignition model.

Thus, the gas phase ignition model has three characteristic times, the sum of which equals the time for ignition, or the ignition delay. These three periods are 1) the thermal induction period, 2) the diffusion induction period, and 3) the chemical induction period. The thermal induction period is the time during which the solid fuel temperature is being raised by external heating to a temperature at which pyrolysis of the fuel begins. During the diffusion induction period, fuel gases are being evolved at the surface and diffusing outward into the oxidant. Finally, during the chemical induction period, the fuel and oxygen react exothermically. Therefore, delaying the onset of ignition implies lengthening one or more of these periods.

The propagation of a flame over a combustible solid flame-spread is an extremely complex process. However, since its rate is readily measured, a large body of experimental data involving flame spread rate has been obtained over the past decade. In spite of these data, there still exists uncertainty as to which physical and chemical parameters exert dominant effects.

Since heat must travel ahead from the flame to unignited material in order to propagate the flame, certain heat transfer modes must be involved. Yet, the relative importance of conduction or convection in the gas phase, conduction in the condensed phase, and radiation in the gas phase with regard to flame spreading rates, is not known even for the simplest cases.

Results from many experimental investigations indicate that the flame spread rate is affected by many parameters such as:

Physical and geometrical parameters:
- orientation of surface
- direction of propagation
- thickness of specimen
- surface roughness
- presence of sharp edges
- initial fuel temperature
- environmental pressure
- velocity of environment
- external radiant flux
- humidity
- specimen size

Chemical parameters:
- composition of solid
- composition of atmosphere

In order to provide general understanding of the relative importance of the various parameters on flame spread phenomena, the effects of some of the physical and chemical conditions listed above are discussed here in some detail.

Angle of Inclination

The results for both polymethyl methacrylate and cellulose acetate indicate that there is little difference between horizontal flame spread rate and vertical downward spread rate over thick specimens.

The much-more-rapid upward spread has not been studied much. The flame spread rate increases with the angle of orientation, by an order of magnitude from the horizontal to vertical orientation.

Thickness of Specimen

Flame spread rates have been found to be independent of fuel bed depths greater than approximately 0.25 cm, but they are inversely proportional to thickness for thinner sheets.

The effective thermal thickness of a fuel bed strongly influences flame spreading phenomena. However, the effects of environmental pressure, oxygen mole fraction, and initial fuel bed temperature on the spread rate differ for thermally thick and thermally thin specimens.

Surface Roughness and Exposed Edges

The physical nature of a polymer can strongly affect the flame spread rate. A flame propagates approximately five times as fast over a smooth horizontal surface where edges are exposed as when the edges are inhibited.

Environmental Pressure

Flame spread rate has been shown to vary with pressure in accordance with the relationship:

$$V \sim P^N,$$

where V = flame spread rate, and P = environmental pressure for a variety of materials over large ranges in pressure as well as at various orientations. For flame spread over the top surface of horizontal thermally thick specimens, and for vertically downward spread over thermally thick specimens, the exponent $N \cong 0.75$. However, for thermally thin specimens and vertical downward spread, $N \cong 0.1$. For other orientations, or for specimens in the transition thickness between thermally thin and thermally thick, N falls somewhere between these two values.

Initial Fuel Temperature

Increasing the initial fuel bed temperature increases the flame spread rate.

Velocity of Environment

The influence of forced convective motion on the flame spread rate is strong.

External Radiant Flux

As might be expected, augmentation of the heat transfer forward from the flame with an external radiant heat flux increases the spreading rate.

Environmental Composition

Both the oxygen mole fraction and inert diluent have been shown to affect the flame spread rate.

Summary of Flame Spread

A number of factors affect the flame spread rate in addition to the chemical composition of the fuel bed itself. Various theories have been proposed to rationalize some of these effects. Models differ widely on the assumptions made concerning the structure of the spreading flame and the mode(s) of heat transfer selected as being important. While all of these models have been successful in predicting some of the experimentally observed effects, each has been criticized for not predicting all effects.

5.2.4 Applications of Fire Dynamics to Test Method Development

In general, our knowledge of fire dynamics is still rudimentary; we are unable to point to examples of its use in test method development. Also, most tests were developed when such knowledge was even less developed, or at times when enormous pressure existed to develop tests in a matter of months, permitting no time for a realistic fire dynamics approach. Even now tests are often introduced in response to a specific disaster rather than as part of a long-range development plan.

However, in a recent case, fire dynamics considerations were used in test development for fire spread down a corridor with flammable floor covering. In this case, corridor fire experiments were performed; the radiation from the ceiling, as well as from the burning gases just below the ceiling, was identified as critical in promoting a "flashover condition" (propagation rates of the order of a foot per second) over the floor covering. Thus, a test is being developed involving radiant heat impinging on a floor-covering sample, with piloted ignition at one end. This test procedure (Flooring Radiant Panel Test) was approved by the ASTM E-5 Task Group on Flooring Materials in June 1975.

The difficulties of modeling a fire by reducing scale arise in several ways: 1) very small fires are laminar while larger fires are turbulent; 2) as far as fluid mechanics is concerned, the ratio of buoyancy forces to viscous forces in the convective flow of fire gases is size-dependent; 3) the radiant emission and self-absorption of the flame are size-dependent; 4) the gas-phase time scale in the fire is shorter for small than for large fires, with possible effect on incomplete combustion products. These formidable difficulties have kept people from having much confidence in fire model test results.

However, it is anticipated that valid modeling procedures may be developed for at least some aspects of fire behavior as the understanding of fire dynamics relevant to situations of interest improves. Then, compensation for any errors or distortions introduced by the modeling might be achieved by varying other parameters such as ambient temperature, pressure, oxygen concentration of ambient radiation.

5.2.5 Conclusions and Recommendations

Conclusions: Fire dynamics is emerging as a scientific discipline. It encompasses chemistry, physics, mathematics, fluid mechanics and heat and mass transfer. The complex dynamic interaction of fundamental processes in fire dynamics coupled with a relatively small funding effort, has led to rather slow progress in quantifying fire behavior. Recommendation: Establish increased program of sufficient magnitude and stability, funded by the government, in the specific area of fire dynamics.

Conclusion: Fire dynamics funding for FY 1977 is estimated to *decrease* 26% below the FY 1976 level, because of federal government budgetary actions. An *increase* of perhaps 8% is needed even to maintain the past level of effort, because of inflation. Recommendations: Support at least 50 additional scientists and engineers (approximately equal to the current effort) over a 10-year period in the specific area of fire dynamics research. This increase would represent an additional cost of approximately $2.5 million per year (approximately 0.02% of the annual cost of fires in the United States, as estimated by the President's Commission on Fire Protection and Control). Bring the program to full strength within the first four years in view of the shortage of qualified engineers and scientists in the appropriate disciplines.

Conclusion: Our current level of understanding of fire dynamics is not sufficiently developed to permit thorough scientific analysis of most fire test methods, fire scenarios or fire modeling procedures. The behavior of a given material in a fire is dependent not only on the properties of the fuel, but also on the fire environment to which it is exposed. Consequently, if test methods are to be meaningful, they must simulate the critical fire dynamic conditions. An understanding of fire dynamics is essential, if the critical conditions are to be identified. Consequently, fire dynamics can be extremely useful in test method selection and development. Valid modeling procedures, based on evolving fire dynamic principles, offer promise for reduction in the dependence on, and the cost of, large scale fire tests. Recommendations: Employ fire dynamics expertise when developing new test methods and for validating and improving existing test methods. Conduct this fire dynamics program primarily in academic, non-profit, and governmental research organizations. The projects should be closely monitored by advisory boards including broad representation by manufacturers, users, and members of standards-setting organizations. The teaching of fire dynamic principles should be included in the training of fire protection engineers, code officials, architects and subway car designers, etc.

5.3 Guidelines for Developing Fire Scenarios

5.3.1 Introduction

Judgment and extrapolation are very important in developing scenarios because only very limited data are available and technological change may occur so rapidly that the time lag between the introduction of new materials, products, or structures and the development of statistically significant accident histories may be unacceptable. It must be remembered that a scenario is a hypothetical failure mode and effect analysis intended to be used as an aid to further study.

In the following discussion of fire scenarios, the physical behavior of the fire is emphasized; the interactions with humans are deemphasized because this study is directed at fire safety via modifying materials and design rather than people.

5.3.2 Ignition Source

Appropriate information about the ignition source is required to characterize it quantitatively because, in many instances, ignition of the target fuel is marginal, i.e., embers fall on a rug but self-extinguish or a blow-torch impinges on a plywood panel but only chars it.

The primary parameters of the ignition source are:

maximum temperature (°C)
energy release rate (cal/sec or watts)
time of application to target (sec)
area in contact (cm^2)

In some cases, details of the heat transfer mechanism from the ignition source to the target fuel must be known because it may be a critical combination of conduction, and radiation. The degree of air motion or turbulence may influence spontaneous ignition of a heated vapor rising from a surface.

The most important single fact to recognize about a potential ignition source is that, for solid polymers which are not readily ignitable, a "strong" ignition source will generally ignite the target while a "weak" one will not.

5.3.3 Ignited Material

The first material to be ignited is important in a scenario because the probability of ignition occurrence depends on the properties of the target material. If the target material is a flammable liquid, its ignitability will depend on whether it is in the form of a stationary pool, a foam, a mist, or a spray.

If the target material were a gas mixture, ignition would depend primarily on the thermodynamic properties of the gas (composition, temperature, pressure).

In most fire scenarios, the target material is *solid.* Ignition of solids results from ignition of the pyrolysis products which are evolved during thermal decomposition of the solid, and occurs in the gas phase above the solid surface.

When considering the chemical composition of a target material, the following factors are especially relevant: 1) basic material may contain small percentages of additives (fire retardants) or impurities, which may have major effects on ignitability; 2) if the material were hygroscopic, like cotton or paper, the initial moisture content will vary over a wide range depending on prefire humidity, with important influence on ignitability; 3) if the material contains several major constituents, as flexible polyvinyl chloride contains a large proportion of platicizer, its ignitability will depend on the more volatile flammable constituent, in this case the plasticizer;

4) the target may be composite in nature, consisting of an outer skin material and an underlying material, either of which may contribute to ignitability.

The thickness and thermal properties of a material are vital in determining the time required to achieve ignition, when a given heat flux is applied to the surface, and they become crucial in the scenario if the heat flux is of relatively short duration. A distinction must be made between "thermally thick" and "thermally thin" materials because the time to ignition for "thermally thick" material is independent of the thickness; it is controlled by the "thermal inertia," which is the product of the thermal conductivity and the heat capacity per unit volume. For a "thermally thin" material, the time to igntion is proportional to the product of thickness and heat capacity per unit volume (fabrics generally fall within this category). Whether the material behaves in a "thermally thick" or "thermally thin" manner depends not only on the thickness, but also on the heating rate, the heating time, and the "thermal diffusivity," which is the ratio of thermal conductivity to heat capacity per unit volume.

In the case of a thin flammable material (carpet, paneling, etc.) in thermal contact with an underlying material, the thermal properties of the underlying material can influence the ignitability by the degree to which the underlying material acts as a heat sink.

The configuration of the target material can also be of great importance. Whereas the foregoing discussion implied a one-dimensional geometry, ignition actually tends to occur more readily in a crevice or fold, or at the edge or corner, rather than in the middle of a flat surface.

5.3.4 Flaming or Smoldering Combustion

Some combustible materials may burn either in a smoldering mode, like a cigarette, or in a flaming mode. A material also may smolder for a certain length of time and then spontaneously burst into flame.

Smoldering is important in that: a) the smoke or gases produced may permit detection of the fire at an early stage; b) the pyrolysis products may be toxic; and c) a transition to flaming, after a long period of smoldering, may produce a very rapidly growing flaming fire because of the preheating of fuel and accumulation of combustible gases during the smoldering period.

Smoldering may continue for a very long time. For example, a barrel of sawdust might smolder for more than 24 hours. Therefore, scenario analysis should consider the possibility of a long time lag between ignition and active flaming as a function of the target materials.

5.3.5 Fire Spread

Unless a person were wearing or sleeping on the originally ignited item, the fire is not apt to do much damage until it has grown by spreading some distance from the point of ignition. The rate of spread is very important in the scenario, because it to fire spread by virtue of its radiation emission or absorption; and, substantial property damage may be caused by smoke or corrosive combustion products.

Optical scattering properties of the smoke depend strongly on particle size as well as concentration, so that vision-obscuring aspects which interfere with escape or firefighting are strongly dependent on the type of combustible and mode of combustion. For example, incomplete burning of polystyrene or rubber produces large soot particles capable of obscuring vision even at low concentrations.

Visibility at floor level will generally be much better than at higher levels in a room, so the possibility of crawling to safety is important.

Toxic Effects

Carbon monoxide is the chief toxicant, according to our present knowledge, but other substances such as acrolein, HCN, HCl, HF and CO_2, which may be present in the smoke can be very hazardous in certain cases and might exhibit synergistic effects.

The critical survivable concentration of toxicant depends on the time of exposure, which, when escape is not possible, in turn depends on the history of the fire.

5.3.7 Flashover

Flashover, a critical transition phase of a fire in a compartment, generally occurs in ventilated compartments, otherwise the fire will tend to smother itself before the flashover stage is reached. Prior to flashover, the rate of the local fire burning in the compartment is determined by the extent of flame spread to that time. After flashover, all flammable contents in the compartment are burning or rapidly pyrolyzing, flames are projecting out the door or window, and the burning rate within the compartment is determined by the rate of ventilation and/or the total exposed fuel area. Flashover often occurs suddenly, within a few seconds, and is characterized by very rapid fire spread throughout the compartment, with flames violently rushing out the door or window.

Whether flashover can occur in a compartment depends on the compartment's size and shape, ventilation available, intensity of the initial fire, as well as the quantity and disposition of secondary fuels. If flashover can occur, the time required for its occurrence will depend not only on the foregoing variables, but also the thermal inertia of the room, especially the ceiling. A fire in a typically furnished room will require five to twenty minutes after flaming ignition to reach flashover.

During the pre-flashover period, the upper portion of the room is filled with hot, smoky, oxygen-deficient combustible gases. The lower portion contains relatively cool, clean air coming from the door or window. At some intermediate height, perhaps two feet under the ceiling, there may be both sufficient oxygen and heat to readily ignite target fuels at that level. Drapes or curtains are examples of materials in this region. Occasionally the hot combustible gases ignite as they reach an adequate source of oxygen resulting in rapid and violent combustion of the major defines the time after ignition when the fire reaches a dangerous size.

Fire may spread either between contiguous fuel elements or by jumping across a gap from the initially ignited material to a nearby combustible item. These two cases are discussed separately as follows:

Fire Spread over the Initially Ignited Material

The rate of flame spread over a solid surface in the horizontal or downward direction is often quite slow, sometimes as little as one inch per minute. However, if the material is "thermally thin," or has been preheated by radiation or convection from hot combustion products, the flame can spread quite rapidly. If the fuel is arranged to permit upward flame propagation, flame spread will occur very rapidly and at a progressively accelerating rate.

Fire spread over a liquid is relatively slow when the liquid is well below its flash point, but possibly a hundred times as rapid if the liquid is above its flash point. For liquids below their flash point, motion within the liquid induced by the fire is important in determining the spread rate.

Flame spread through a premixed gas may vary from a foot per second to hundreds of feet per second, depending on conditions for generating turbulence by the expansion associated with combustion. For some conditions of confinement, detonations moving at velocities of several thousand feet per second occur.

Fire Spread to Secondary Material

When originally burning material is separated by a gap from the nearest secondary combustible, and the flame does not impinge directly on this secondary material, the fire will die out after the original material is consumed, unless by some mode the fire can spread across the gap as discussed below:

A fire may radiate directly on the target or convectively heat the ceiling and upper walls, which then radiate onto the target. Alternatively, hot smoke gases accumulating under the ceiling may radiate onto the target; or, hot combustible gases accumulating under the ceiling may ignite and radiate onto the target.

In any event, radiation preheats the secondary material until it pyrolyzes, emitting flammable vapors. At this point two possibilities exist. Either the secondary surface may ignite, or a sufficient concentration of flammable vapor mixture accumulates to permit the original flame to spread through it to the secondary material.

5.3.6 Evolution of Smoke and Toxic Gases

Smoke and toxic gases are important to the fire scenario because they not only may provide means for the early detection of fire, but also may interfere with visibility, escape or firefighting. They have psychological and physiological effects on humans, including confused thinking, incapacitation, and death. In a majority of cases, deaths in building fires are a result of the toxic combustion products present rather than as a result of the heat and flames. Furthermore, smoke may contribute importance in flashover. Thus, the infrared emission, absorption, and reflection characteristics of objects and smoke in the compartment are highly relevant.

5.3.8 Spread to Adjacent Compartments and Catastrophic Failure

Fire resistant compartmented buildings are designed with the expectation that a fire in any one compartment will be confined by the building structure itself so that either the fire is extinguished or the fuel is exhausted before the fire breaks through to adjacent compartments. Interior partitions, fire doors, etc., are subject to building codes specifying that the partition must maintain its integrity for an appropriate time. Thus, the scenario should include information on the fire endurance rating of the relevant structural elements.

If the fire compartment has openings to other sections of the structure, such as open doorways, ventilating ducts, improperly firestopped or inadequately sealed openings in walls, etc., these would constitute elements in the scenario. Even if the fire were confined to the compartment of its origin, the spread of smoke and toxic gases through the structure could have catastrophic effects. If the fire were capable of heating structural elements of the building to failure (steel above 538°C [1000°F]), the structure might collapse. Thus, the thickness and integrity of insulation on structural elements also become important to the fire scenario. Local structural collapse within a sprinklered building may cause breakage of the sprinkler piping and consequent escalation of the fire damage.

5.3.9 Spread to Other Structures

Where a structure becomes completely involved with fire, a finite probability exists that adjacent structures will ignite. Ultimately, a conflagration involving a large area may result. Such fire propagation could occur either by radiation or firebrands.

Potentially critical factors in the fire scenario are: magnitude and direction of the wind, separation distance between structures; flammability of roofing material such as wood shingles, ignitability by radiation of curtains inside windows facing fire, combustible trash in alleys between buildings, and propulsion of burning debris after building collapse or explosion.

Since the spread of fire to other structures occurs at a sufficiently late stage of a fire, firefighters will usually be present. Their tactics in wetting down adjacent buildings are extremely valuable in preventing spread to such structures. Conversely, if the fire were simultaneously burning in many areas, as could be the case for a brush fire or fire caused by civil disorders or military incendiary attack, firefighting will probably be inadequate, and the degree of spread will depend on the intrinsic "hardness" of the structures involved.

5.4 Guidelines for Analysis of Fire Scenarios

5.4.1 Introduction

A scenario should cover as many as possible of the following points.

(a) The pre-fire situation.
(b) The source of the ignition energy should be identified and described in quantitative terms.
(c) The first material ignited should be identified and characterized as to chemical and physical properties.
(d) Other fuel materials that play a significant role in the growth of the fire should be identified and described.
(e) The path and mechanisms of fire growth should be determined. Particular attention should be given to fuel element location and orientation, ventilation, compartmentation, and other factors that affect fire spread.
(f) The possible role of smoke and toxic gases in detection, fire spread and casualty production should be determined.
(g) The possibility of smoldering combustion as a factor in the fire incident, e.g., as a cause of re-ignition, should be considered.
(h) The means of detection, the time of detection, and the state of the fire at the time of detection should be described.
(i) Defensive actions should be noted and their effects on the fire, on the occupants, and on other factors should be described.
(j) Interactions between the occupants of the building or vehicle and the fire should be detailed.
(k) The time sequence of events, from the first occurrence of the ignition energy flux to the final resoltuion of the fire incident, should be established.

The scenario should permit generalization from the particular incident described. It should provide a basis for exploration of alternative paths of fire initiation and growth and for analysis or the effect on fire safety performance of changes in materials, design, and operating procedures.

5.4.2 Questions on Prevention and Control

While the analysis of any fire scenario might be accomplished in several ways, an effective procedure is to ask questions about each essential fire scenario element. The answers to such questions should suggest means for fire prevention and control.

Prefire Situation

1. Were existing codes adhered to?
2. Were materials installed properly?

Ignition Source

1. In as much detail as possible, what was the ignition source?
2. For how long was it in contact with the ignited material prior to flaming ignition?
3. Could the ignition source be eliminated?

Ignited Material

1. What was the originally ignited material?
2. What was the application of the material (e.g., drape, rug, cushion, etc.)?
3. How was it located relative to ceiling and nearest wall?
4. What were the ventilation conditions in the room?
5. What was the relative humidity during the 48-hour period before the fire?
6. Did melting and dripping of the ignited material occur?
7. Did the ignited material collapse, fall over, etc.? If so, what effect did it have on the fire?
8. Once the ignited object is fully involved, is it possible to quantitatively estimate its heat release rate?
9. Are there other materials that could have been employed in this application which would not have ignited under the same exposure conditions?
10. Do flammability tests on materials intended for this application adequately measure ignition resistance to this level of ignition source? Should they?

Flaming or Smoldering Combustion

1. Is is known if smoldering preceded flaming?
2. Can the time-dependent concentration of smoke and toxic gases arriving at a strategic location some distance from the fire be estimated?

Fire Spread

1. How long did it take for the first ignited object to become fully involved?
2. How was flame spread influenced by sudden events, such as breaking of windows, opening of doors, melting and dropping of curtains, spattering of burning droplets, etc.?
3. Did one or two materials significantly control the fire spread rate?

Smoke and Toxic Gases

Automatic Detection

1. If a smoke detector were present, how was it located relative to the fire? Did it respond as would be expected?
2. Supposing no smoke detector were present, how much sooner would the fire have been detected if a smoke detector had been present in a logical location? Would detection have been soon enough to make a crucial difference?

Visibility

1. Was visibility obscured in an escape route?

Toxic Effects

1. Were victims affected by toxic substance?
2. What toxic substances caused death (autopsy)?

3. Did toxic substances interfere with victims' escapes by promoting confused thinking or decision making?
4. Could these toxic substances be attributed to any one material?

Extinguishment

1. How large was the fire when first detected? What were visibility conditions at that time?
2. How much time elapsed between detection and attempted extinguishment?
3. How large was the fire when extinguishment was attempted?
4. What was the estinguishment technique and how successful was it?
5. If automatic sprinklers had been present, how much sooner would they have been expected to control the fire, and how much less might the loss have been?

Flashover

1. Did flashover occur? How long after ignition? How long after detection?
2. Can crucial elements in the fire growth and spread process be identified in relation to flashover?

Postflashover

1. Did fire spread beyond the initial compartment? How? Was door open?
2. How was the ventilation system involved in fire spread?
3. Did walls, fire doors, etc., fail? If so, after how long?
4. Did fire spread to floor above? By what mechanism?
5. Did structural collapse occur? Was it due to code violations?

Spread to Other Structures

1. Did other structures ignite? How far away were they? What material first ignited?
2. Was radiation responsible? Were firebrands involved? Was there direct flame contact?
3. What were wind conditions?
4. Did firefighters attempt to protect exposed structures? How soon before spread occurred?

5.5 Spectrum of Fire Scenarios

5.5.1 Introduction

A few brief scenarios, based on real fire incidents in which polymeric materials play a significant role, are presented to illustrate the diversity of fires and show commonality that permits a scientific approach to fire safety. In these scenarios, attention is particularly directed to the following areas; prefire environment, ignition source, material first ignited, other significant materials involved, fire dynamics, method of detection, extinguishment, and extent of loss.

5.5.2 Aircraft Fire Scenario

• A trans-Atlantic jet was approximately a half hour from its destination when a passenger went to the rear of the plane and opened the lavatory door. He was greeted by a wave of white smoke. He quickly closed the door and called the attendant. The attendant reported the situation to the captain who sent the flight engineer back to investigate, at the same time requesting permission for emergency descent. The flight engineer found the rear of the cabin filling with black smoke. He instructed the cabin steward to attack the fire with portable CO_2 extinguishers while he increased the air flow to the cabin to keep the smoke in the rear of the airplane. The fire extinguishers were ineffective. Within three minutes of the discovery of the fire, the smoke had reached the cockpit and was interfering with visibility. The cockpit windows were opened, making visibility possible, and the pilot prepared for an emergency landing. The aircraft came to rest in a field with the fuselage practically intact, but a fire broke out in the cabin and spread through the cabin in seconds. There were only five survivors. It is believed that most of the passengers were overcome by smoke as well as gas and would have been unable to evacuate the plane without assistance even if the fire had not intervened to prevent rescue operations. The cause of the fire could not be determined, but a discarded cigarette in the lavatory trash container was suspected.

• A jumbo jet, having made a cross country flight, was given a brief post flight inspection and secured for departure on the folloiwng day. Approximately four hours later an airline employee smelled smoke next to the aircraft passanger ramp. Upon boarding the plane he noticed smoke emanating from the rear of the cabin area. He immediately reported the fire to the airport fire department. They responded quickly and found fire burning through the fuselage. An attack was made, and it was realized that additional aid was needed. This was requested. From the initial reporting of the incident to burn through of the cabin area fuselage, approximately 20 minutes had elapsed. The entire cabin area was destroyed. Post-fire inspection revealed that the ignition was caused by a short in an electric outlet in the lavatory attributed to routine maintenance, which ignited paper toweling in a waste container. The fire then traveled up, between the fuselage frames to the overhead ceiling area, and then forward. Burned wiring bundles indicated the intensity of the fire above the ceiling panels. Besides paper toweling and tissues, combustibles contributing to the fire included acrylonitrile butadiene styrene (ABS) or vinyl-type thermoplastics, wiring insulation, wood frames for the vertical and ceiling panels, and the neoprene/nylon vapor barrier covering the insulation blankets. The fire was finally extinguished by the joint efforts of the airport and municipal fire departments.

5.5.3 Building Fire Scenarios

• Two couples had been playing cards in the downstairs recreation room of a two story frame house. About midnight, the visiting couple left and the residents went to bed in an upstairs bedroom. At approximately 2:00 A.M. a neighbor saw flames at the recreation room window and called the fire department. He attempted to rouse the occupants of the burning house, but received no response. The fire department responded within five minutes and found the recreation room fully involved. The fire, which had not spread beyond the recreation room, was quickly extinguished but the occupants were found in their bedroom, both dead. The man was lying back on the bed with his feet on the floor; the woman was at the bedroom door. Smoke stains on the wall indicated that smoke was carried into the upper part of the room by the air conditioning system and accumulated down to a level of four feet from the floor. Examination of the recreation room revealed that an upholstered sofa was almost completely destroyed, draperies at the windows had burned igniting an adjacent plywood panelled wall, and a mineral acoustic tile ceiling was badly discolored, but had not contributed to the fire. The wall-to-wall carpeting was burned in the neighborhood of the sofa and draperies, but had not contributed to the spread of the fire. It is probable that a discarded cigarette, left behind at the end of the card game, ignited the sofa, which smoldered for some time, producing large quantities of smoke and toxic gas before bursting into flame. The limited extent of fire damage indicated that flaming combustion had been in progress for only a short time before the fire was extinguished. Death was attributed to carbon monoxide poisoning.

• A fire started at about 1:30 A.M. in a sixth floor apartment of a twelve story apartment building for the elderly. The cause was unknown, but the occupant of the apartment was known to be a chain smoker. The occupant got out of the apartment, but collapsed in the corridor outside, leaving the apartment door open. The body was found near the open door after the fire was extinguished. The building automatic smoke alarm alerted the attendant on duty at the front desk. He took his elevator to the sixth floor to investigate. His body was found later in the sixth floor elevator lobby just in front of the elevator. At 1:45 the fire was reported to the fire department by telephone from a neighboring apartment building. When the firefighters arrived, they found that the fire had spread from the room of origin down the corridor over the carpet. A one and one half inch gap for ventilation under the doors of other apartments not only allowed limited fire spread over the carpet into those apartments, but also permitted smoke and gas penetration. In addition to the occupant of the room of origin and the building attendant, five other persons died of carbon monoxide poisoning in their apartment on the sixth floor. Fire resistant construction prevented structural damage and spread of the fire to other floors.

• Two construction workers were using a torch to cut holes in the interior wall of an unsprinklered refrigerated warehouse to install a new conveyor. The wall was of steel construction with three inches of sprayed-on plastic foam insulation on the inside. Shortly after the work was started, the workmen observed fire in the interior of the warehouse around the hole they had made. They entered the warehouse through a door and found the insulation burning. One went to call for help while the other attempted to fight the fire with a portable fire extinguisher. When the first one returned to the scene, black smoke was pouring from the door. He was unable to enter the warehouse or locate his companion. After the fire department arrived, firemen wearing self-contained breathing apparatus quickly extinguished the fire. The body of the workman who had tried to fight the fire was found near the door. He was unburned. A portion of the insulation had been burned from the wall, but there was no structural damage to the building. Restricted ventilation had limited the growth of the fire.

- A plant that manufactured expanded polystyrene egg cartons from polystyrene pellets used a process that involved extrusion with a blowing agent, followed by thermal forming in presses. The egg cartons were packed 200 per polyethylene bag, with 18 bags stacked on a 4′ X 4′ piece of cardboard, making a 6′ high stack. The storage array was 3 stacks high. or 18′ to the top. The storage area was sprinklered at the ceiling level. Ignition was caused by an overhead lighting fixture in contact with the stored bags. The fire burned with such intensity that sprinklers, at 0.4 gallons per minute per square foot, were unable to control the fire. The fire department arrived in ten minutes, but the dense smoke hampered its operations. Three firemen were overcome by fumes from the burning plastic. After three hours of burning under conditions of restricted ventilation, a large portion of the roof over the fire collapsed. Rapid growth of the fire followed, driving the firemen from the building. Molten polystyrene covered a 10,000 square foot area to a depth of 5 inches on the floor. Damage to the building, machinery, and stock totaled $1.9 million. Production was interrupted for 8 months, at a loss of $1.5 million.
- A group of teenage inmates of a state "training school" barricaded themselves in a recreation room in protest against disciplinary policies. The outside exit to the room was locked and windows were barred. The room was of non-combustible construction. The inmates piled furniture, including chairs and sofas upholstered with polymeric foam material, against the door leading to the rest of the building. One of the inmates set fire to the furniture to attract attention. They then tried to suppress the fire by smothering and beating it out with materials at hand, but were unsuccessful. The fire grew rapidly in the pile of furntiure, filling the room with smoke and gas. The fire alarm sounded and help arrived promptly, but, by the time the door was forced open, the room was completely filled with smoke. The fire department quickly extinguished the fire, which was confined to the furniture. One inmate was dead from inhalation of toxic gases and the others required hospitalization. There was no significant structural damage to the building.

5.5.4 Vehicle Fire Scenarios

- Children were left in a locked automobile while the parents went shopping.

The children, playing in the back seat, found a pack of matches. The upholstery materials in the back seat became ignited, and then the foamed plastic padding material became involved. The children died of burns and smoke inhalation before the fire department arrived. Rescue was attempted by several passersby, but the heat was too intense.

- A bus driver had just parked his new bus at a turn-around spot after unloading his passengers approximately two blocks before. He smelled smoke, and upon turning in his seat saw smoke coming from the rear of the vehicle. He attempted to extinguish the fire with a portable extinguisher, but was driven back by the intense smoke. The foamed plastic seating was the initial material observed to be on fire. Shortly thereafter, the vinyl wall coverings and synthetic plastic windows became involved. Flashover occurred in less than three minutes after discovery of the fire and the entire bus interior became involved. The interior of the bus was destroyed in less than five minutes, before the fire department could extinguish the fire. Fire officials investigated the fire and concluded that arson was its probable cause.
- As a subway train approached an underground station, a blown rubber radial tire caused intermittent contact with the live third rail. This intermittent arcing occurred until the tire and hydraulic lines in the undercarriage caught fire. After these components ignited, the fire penetrated the plywood undercarriage and proceeded to burn the car interior, which consisted of plastic interior finish and foamed plastic seating. Passengers had been removed prior to the burn-through and a trainman attempted to extinguish the fire by directing a portable carbon dioxide extinguisher at the flames. His efforts were unsuccessful and the fire was soon out of control. Fire fighters were unable to approach the fire for over two hours due to the smoke and intense heat. Four cars out of ten were completely destroyed.
- A ship was alongside a tender for upkeep and minor repairs. Damaged and worn out plastic foam mattresses were being replaced. Several of these items were briefly stacked against a metal bulkhead until the hatch could be cleared for their removal. An engineering technician was using a torch to cut metal support clips from the opposite side of the bulkhead. He had properly requested, received and instructed a fire watch on his side of the bulkhead. No fire watch was ordered for the other side because there were no openings in the bulkhead, the cutting torch would not penetrate, and the bulkhead was presumably devoid of combustibles. Yet, heat from the torch, transmitted through the metal bulkhead, ignited a mattress resting against it. The fire was discovered when smoke was seen pouring from an open hatch. Heavy black smoke hindered firefighting efforts. Fortunately, there were no combustibles other than the mattresses present to spread the fire. The fire was extinguished without serious structural damage, but with extensive smoke damage to the compartment of origin and adjoining compartments. Several crew members suffered from smoke and gas inhalation, but recovered without permanent injury.

5.5.5 Mines and Bunkers Fire Scenarios

• A mine's electrical system had recently been renovated and new high voltage power lines were fitted underground. The previous terminal box, specifically designed to be explosion proof, was utilized for the new wattage. Unfortunately, the PVC covered cable that was joined in this large 3′ X 4′ X 4′ junction box, required more ventilation to carry away the heat generated by the increased power supply. Constant overheating and cooling caused the insulation material to break down to form an explosive gas mixture in the box. Subsequent high voltage arcing caused the heavy metal terminal box to explode with tremendous force, throwing chunks of metal 400′ down the shaft. One large section of the box severed emergency power cabling putting sources of light and ventilation in the shaft out of action. The explosion started a fire, which spread to a rubber covered conveyor belt. To extinguish the fire required extensive efforts over a two day period. The entire 400′ length of belt required replacement.

• Fire retardant-treated urethane foam was being installed as ventilation control in a coal mine shaft. The material was being foamed in place and its overall thickness was a nominal 6″. The equipment operator inadvertently utilized a mixture too rich in catalyst. After completing the block, he left the area and removed his equipment. Four hours later the excessive catalyst caused the urethane to self-heat and burst into flame. The fire was detected by smoke rising out of the ventilator shafts and a quick response was made. Teams of experienced fire fighters made their way to the entrance of the shaft and utilized the prevailing ventilation to speed the delivery of High Expansion Foam, which extinguished the fire.

5.5.6 Tents, Recreation Vehicles, Mobile Home Fire Scenarios

• While a housewife was preparing a meal in the kitchen of a mobile home, grease in a frying pan on the stove caught fire. She called for help, but, by the time her husband could respond, the foam plastic cabinet above the stove was involved. He brought a garden hose from outside, but by this time the flames were spreading over the plywood interior finish and he was unable to enter because of the heat and smoke. The fire department, called by a neighbor, responded within five minutes, but, by this time the trailer was completely involved, resulting in total loss.

• Two nine-year olds were sleeping in a pup tent in their back yard. They had built a fire near the open end of the tent earlier and had gone to bed when it burned down to glowing coals. During the night, a gust of wind blew a spark against the tent, starting a smoldering fire, which was fanned by wind and burst into flame. The fire spread rapidly over the water repellant treated fabric. The parents, sleeping in the house, were awakened by the children's cries. One child escaped through the burning open end of the tent and received severe burns when his clothing caught fire. The second child attempted to get out through the closed end of the tent away from the fire, but was trapped under the collapsing tent and died of burns. There was no spread of fire to other structures.

5.6 Analysis of Specific Fire Scenarios

5.6.1 Introduction

To illustrate the fire scenario approach to augment fire safety, two generalized fire scenarios are developed; one deals with an apartment fire, the other with an inflight aircraft fire. These scenarios, which include the essential fire elements (e.g., ignition sources, first material ignited, etc.), permit generalization from the particular incident described. Moreover, these scenarios are analyzed to identify hazards, suggest opportunities for fire prevention, and direct attention towards methods for control. The analyses question the adequacy of specific critical materials, design approaches and regulatory codes and procedures; although the committee has excluded "review for adequacy of specific codes" from its responsibilities.

The specific goal of this paragraph is to demonstrate that fire scenario development and analysis constitute a productive methodology for improving the selection and use of polymeric materials to increase fire safety in our increasingly complex systems.

5.6.2 Apartment Fire

A fire started in a lounge chair in the northwest corner of Mr. and Mrs. John Doe's living room. The Does lived on the second floor of a recently constructed three floor condominium apartment building. Each floor contained four similar two bedroom apartments clustered about a stairwell. Construction, typically of wood, met local and FHA codes including those for fire walls, doors, etc. One course of brick veneer was applied to the outer walls. Access to the building was provided through a front hallway to an open stairwell. The second floor hallway had steps leading up and down. In addition to the steps leading down, the third floor hallway contained a ladder leading to an attic-type partial floor above it. Since only three floors above ground level were occupied, there were no fire escapes or other provision for secondary egress.

The furnishings, made entirely or partially of polymeric materials, included furniture, rugs, draperies, wall coverings, etc. The plumbing system was constructed of plastic pipe; the combination tub-shower stall was also made of plastic.

Mr. Doe had been watching the Monday night football game after enjoying a late dinner. Two cocktails, a pleasant dinner and a liqueur made the first half of the game more enjoyable. At halftime, Mr. Doe had another drink followed by a cigarette and then a second cigarette. Mr. Doe drowsed off while smoking the second cigarette; he was awakened by the noise of the crowd cheering a touchdown and decided to turn off his television set and go to bed.

The second cigarette had fallen from his hand into the chair. It smoldered there for 45 minutes before a flame appeared. The flame grew, going up the back of the chair and through the cushion. Heated pyrolysis and combustion gases rose to the 8 foot ceiling. Burning intensity increased rapidly, spreading smoke and hot gases into other rooms. Flashover occurred in 8 minutes (from flaming ignition) propagating flames into the bedrooms, hallway, bath and kitchen. Hot combustion products and smoke poured through the front door, opened by Mrs. Doe during escape, into the center hallway and up the stairwell. Plastic fixtures in the bathroom quickly pyrolyzed, then burned. Smoke and flames followed the plastic plumbing system into the floor above, breaking out into the bathroom. The fire continued to grow on the second and third floors. It burned more intensely on the second floor, engulfing structural elements including the main stairwell. Firemen were called by an occupant of another apartment at this point, arriving simultaneously with the flashover of the upper stairwell. They rescued some of the occupants by ladders through windows, contained the blaze in 15 minutes, and put out all visible flames in less than two hours. Mop-up operations continued for several hours.

The Doe's apartment, and the one directly above it, were gutted. Downwind apartments on the second and third floors were heavily damaged (the wind was about 10 mph). Lesser damage was suffered by immediately adjacent apartments, but the ground floor, while undamaged by flames, did experience water damage.

John Doe didn't make it from the bedroom. His wife, awakened just before flashover in the living room, picked up their child and ran out the front hallway door, leaving it open.

Two elderly persons living on the third floor, over the Doe's apartment, died of smoke inhalation and/or burns. In the downwind third floor apartment, all four occupants, although affected by smoke, were rescued by firemen, as were the occupants of the adjacent apartment. Ground floor occupants escaped unharmed.

Since this scenario is plausible, its analysis in detail should reveal corrective measures applicable to similar occupancies and perhaps to general fire situations.

Ignition Source

The ignition source was a cigarette that fell into the overstuffed chair.

Not much can be done to reduce the ignition capabilities of today's carelessly discarded cigarettes. A type of cigarette that goes out after a short time if not smoked exists; it should be further developed since cigarettes are a frequent cause of fires. Today's cigarette requires favorable conditions before it will ignite other objects. Many carelessly discarded cigarettes go out without igniting other materials. A continuing education program emphasizing the dangers of smoking while half asleep, in bed, or in a semi-intoxicated state, offers some promise; but major efforts should be directed toward countering the ignition source since this approach offers great potential for reducing fire incidents.

A cigarette is a serious ignition hazard. Because of its self-perpetuating glowing (not flaming) combustion, it can remain undetected in a smoker's environment for relatively long periods of time. This allows the fire a long period to develop and become deep seated. During such long induction periods, large quantities of lethal gases may be produced. Fires from other ignition sources, such as matches and lighters, are usually quickly detected and countermeasures can be started promptly.

Other ignition sources that could have initiated this type of fire (i.e., slowly developing) include "instant-on" televisions, electrical heaters, fireplaces, and overloaded or damaged wiring.

Ignited Material

The cigarette was accidentally dropped into the space between the seat cushion and the arm of a typically upholstered overstuffed lounge chair. The cotton covering material smoldered at first, allowing the heat from the glowing cigarette (800°C) to contact the urethane foam cushion.

Within the present state-of-the-art of materials, there is much that can be done to improve the fire resistance of chair coverings and cushions. In fact, current development programs have already produced cushion materials and covering materials of substantially higher resistance to ignition and flame spread. NBS is developing new testing methods that measure the response of materials to cigarette ignition sources. Moreover, the Consumer Product Safety Commission is considering the promulgation of a mandatory standard which would eliminate easily-ignited upholstered furniture from the market place. Completion of this test program as well as the application of new materials and designs as appropriate would promote improved fire safety.

Fire safety for apartment occupants could be markedly increased by the installation of an inexpensive smoke detector, which would provide a warning during an early stage of the fire, probably during the smoldering period.

If the ignited material had been discovered earlier, the fire could have been contained rather easily by cutting out the affected portion and removing it, or the entire chair, from the building. A smoke detector and a properly sized fire extinguisher should be a part of the furnishings of every home.

The analysis might also consider other likely ignition sources and their locations in the apartment. Kitchens produce many fires from grease appliances. Smoke detectors in kitchens are not useful, because of fire alarms from cooking smoke. However, since kitchens are usually occupied when flames erupt, it is much better to have a fire extinguisher on hand. Kitchen trash containers often receive cigarette ashes and butts. Trash containers should be metal rather than combustible material. In the past, bathrooms have been relatively fire safe, but the new plastic tubs and showers can contribute significantly to fire growth. Bedrooms have several potential fire sources, including people who smoke in bed, children who may play with matches, appliances, wiring, etc. A hall smoke detector would provide some warning of fires from all these locations. Individual families, when assessing their living habits and accommodations, could develop priority requirements for the acquisition of fire detection and control equipment.

Smoldering and Flaming Combustion

Although the cigarette, upholstery material and urethane cushion smoldered for about 45 minutes, they produced only small amounts of smoke and gas. Then, burning characteristics changed to full flaming combustion, rapidly engulfing the chair arm and cushion. Large quantities of hot smoke and gas developed and rose to the ceiling.

Even with the given ignition source characteristics and system geometry, current covering and cushioning materials as well as designs, when properly employed, can significantly increase fire safety. Additionally, some exotic "space age" materials that are extremely fire resistant could be brought into common use through continued efforts to reduce their costs.

Control of a fire by apartment dwellers following the transition to flaming combustion is difficult. If the fire had not been noticed until the flame broke out, the safest procedure would have been to leave the apartment, notify the other occupants of the building and call the fire department. This course of action was not available to the sleeping Doe family.

As flames broke out in the back of the chair, the polymeric window draperies caught fire. The fire spread to the wall paneling and nearby coffee table made of structural foam. Soon, the entire corner of the living room was blazing; producing copious quantities of hot smoke and noxious gases that rapidly filled the small plenum overhead and began to flow through the top of the living room entrance way into the hall and other rooms with open doors. Radiation from the hot gases and ceiling heated the other furnishings in the room to their ignition points and shortly thereafter flashover engulfed the living room.

Review of this segment of the fire chain revealed several excellent possibilities for increasing fire resistance:

1) Replacement of the draperies with drapes made of more fire resistant material, e.g. fiberglass.
2) Replacement of the paneling the gypsum wall board.
3) Replacement of the coffee table and other plastic furniture with wooden furniture, or more fire resistant plastic furniture.

Control of the fire at this stage demanded professional fire fighters and their equipment, but at this time in our scenario they had not been summoned.

Evolution of Smoke and Toxic Gases

Smoke and toxic gases evolved in each of the phases described earlier. Smoldering combustion generally produces more CO and less heat than does flaming combustion; smoke quantities vary considerably.

Smoke detectors are available in a wide range of capabilities and price. Relatively inexpensive, but quite effective, detectors are available. Heat detectors are also available at a wide range of price, but they are better suited to property protection than to life-safety applications, since they are less *rapid in their response.* Reliable and inexpensive detection of the toxic gases, principally CO, is very difficult.

Visibility is markedly affected by smoke. Smoke obscures or obliterates familiar landmarks and causes some distortion; it irritates the nose, mouth and throat. Toxic gases and smoke affect the eyes (watering) and reduce visual efficiency in many other ways. Toxic gases have severe effects on the brain, nervous system, respiratory system and heart. Although it has not yet been quantitatively determined, there is some qualitative evidence that combinations of toxic gases interact on humans in a way such as to multiply their individual effects. The toxic gases generated in this scenario were sufficient to produce the demise of all members of the Doe family.

In a fire, thermal burns are occasionally the cause of death. More frequently, as in our fire scenario, thermal burns on the outer skin or in the respiratory system are not the primary cause of fatalities. Smoke and toxic gases incapacitated the victims, preventing their escape. The burns suffered by Mr. Doe were a secondary effect.

Flashover

Shortly after flaming combustion engulfed part of the living room, the superheated pyrolysis products filled the upper levels of the hallway, kitchen, dining area, and bedroom through the open doors. Flashover occurred initially in the living room. Mrs. Doe awakened at this point, grabbed her child and was just able to leave the apartment. Smoke and gases rapidly filled the bedrooms, flames moved laterally along the upper walls supported by new air coming in at the bottom of the main entrance doorway. Hot gases filled the upper part of the bathroom and pyrolyzed the plastic shower walls that fed still more hot gases. Flashover followed shortly thereafter and the plastic tub burst into flame sending hot gases up the plumbing shaft into the bathroom of the apartment above. One by one the bedrooms flashed over and were engulfed by flames.

As noted earlier, once conditions for flashover develop, full scale firefighting efforts are required to stop the process and protect adjacent structures.

Spread to Adjacent Compartments

Hot smoke and gases flowed from the Doe's apartment into the hall and stairwell, rapidly filling the third level. Flames exited the Doe's door and ignited the walls of the hallway; flashover occurred on the third level in a matter of minutes, gradually engulfing the hallway in flames from the second to the third floor. Simultaneously the fire broke through the bathrooms and the living room into the third floor directly above. The elderly occupants of this apartment were unable to escape. On the other side of the condominium building, occupants were unaware of the fire until the stairwell flashed over. The third story occupants were rescued by firemen, then the second story occupants were rescued. All first story occupants, aroused by flashover noises and the fire department's arrival, were rescued or exited unassisted. Meanwhile, the fire spread through structural walls, fed by foamed polymeric insulation.

No fire escapes were required by codes, therefore, none were provided. None of the third floor occupants had installed portable ladders for escape.

Fortunately the windows and storm sash could be opened for escape. Sealed windows and certain casement, jalousie or awning type windows make escape difficult or impossible. Special escape routes planned by the architect may be required under such circumstances.

Full sprinkler installation in all portions of the condominium unit would clearly add to fire safety. Such installations may be too costly for some types of housing. A sprinkler system for the common use stairwell, and particularly for the furnace room, would offer major improvement of escape possibilities at a much lower cost.

Since at this phase of the fire chain, control and extinguishment requires fire professionals and their equipment, the analysis does not reveal substantial opportunity for improving fire safety except through optimum use of improved materials, self-closing doors, early detection of the fire, and prompt notification to fire fighters.

Summary

Analysis of this hypothetical scenario led to several recommendations that, if implemented, would be effective in reducing the adverse effects of this type of fire, at relatively low cost, for the specific types of dwelling units examined. These recommendations include:

1) Require upholstered furniture to have a covering material — foam cushion combination resistant to a cigarette ignition source.
2) Each apartment and common hallways should be equipped with smoke detectors for early warning.
3) Sprinklers should be installed in hallways and stairwells.
4) Building design and construction practices should include provisions for fire safety. Education of architects, engineers and builders in fire safety would be particularly beneficial.
5) Develop an acceptable cigarette which will self extinguish after a short time if not smoked.

5.6.3 In-Flight Aircraft Fire

In this hypothetical scenario a transatlantic commercial jet, nearing the end of its flight, was approximately 30 minutes from its destination. A passenger, upon entering a closed lavatory at the rear of the aircraft, encountered a wave of white smoke. He quickly closed the lavatory door and called for the attendants.

The first steward went into the cockpit and reported smoke and fire in the aft lavatory. A cabin crew member discharged two fire extinguishers into the lavatory, but smoke continued to increase and become darker.

The captain reported the fire to ground control and requested emergency descent. At the same time, he ordered the cabin depressurized and sent the flight engineer back to analyze the situation. The flight engineer went to the rear of the aircraft taking with him a CO_2 fire extinguisher bottle. When the flight engineer saw black smoke filling the area behind the last row seats, he handed the extinguisher to a steward and quickly went back to the cockpit to report to the captain. As the aircraft was descending, the flight engineer increased the airflow to the cabin to keep the smoke in the rear of the airplane.

Soon thereafter, a steward came into the cockpit reporting that the cabin was half filled with smoke and passengers were being affected. The captain ordered an overwing emergency window removed. A steward, equipped with an O_2 bottle and a full face mask, tried unsuccessfully to comply with that order.

Approximately 3 minutes after the first report of smoke and fire in the lavatory, smoke reached the cockpit reducing visibility so that the pilots could neither see the instruments nor through the windshield. Both pilots opened their sliding windows. Visibility was made possible through the open windows as the flight continued.

The captain decided to land as soon as possible and landed the aircraft soon thereafter in an open field. During the landing both main landing gears broke off. The fuselage came to rest practically intact. After the aircraft came to rest, the fire, already in progress within the fuselage, broke out through the top of the cabin in front of the vertical fin. This fire consumed virtually the entire fuselage interior. Only one passenger and four crew members survived.

Analysis of the Aircraft In-Flight Fire

The aircraft was constructed by an established manufacturer, competent in every respect, who complied fully with applicable regulations. Operating procedures were also in accordance with regulations. The crew was well trained and fully experienced. The aircraft had been properly serviced and loaded.

Ignition Source

The fire started from a cigarette dropped into the paper towel disposal slot. It landed on a dry paper towel and a film wrapper from a pack of cigarettes.

Despite a ban on smoking in lavatories, some persons ignore it. They drop ashes into the wash basin and cigarette butts (hopefully extinguished) into waste containers out of ignorance or to conceal the violation of the no-smoking rule. Effective enforcement, possibly using smoke detectors, would eliminate this problem.

Ignited Material

The material ignited by the cigarette was the cigarette package wrapper and paper toweling discarded through the waste disposal slot into the waste bin. Either could have supported ignition, although waste paper toweling often smolders and extinguishes. Since paper is a proper product for the dispenser, disposal arrangements should be fire hardened. The waste bin was plastic. One step in fire safety would be to replace the plastic bin with a non-combustible bin.

Smoldering and Flaming Combustion

The ignited cigarette package wrapper flamed. The discarded paper toweling smoldered and then flamed involving more paper. Soon the plastic trash bin began to pyrolyze and flame appeared, travelling upward, involving the plastic paneling of the lavatory framework.

The waste bin could contain all kinds of combustible contents other than used toweling. Thus, either smoldering or flaming combustion could occur even from a low heat flux ignition source. Corrective measures require more than just fire hardening the trash container. Although system design *per se* is not within the committee's expertise, it appears that water application to the trash, or occasional inerting by nitrogen or carbon dioxide, offers worthwhile safety advantages. Another safety precaution would be the presence of a small self-actuating fire extinguisher.

Flame Spread

The fire spread up the plastic paneling, penetrating the lavatory overhead, and began to work itself laterally, fore and aft. Smoke filled the adjacent lavatories, moving forward in the overhead plenum and into the passenger cabin. As the fire grew, additional smoke and gases poured into the cabin and moved forward into the cockpit, obscuring the vision of the crew.

Once the waste bin and lavatory started flaming combustion, there was little the crew could do to extinguish the burning materials with the fire extinguishers (CO_2 portables). The openings above, below, and between lavatories facilitated the fire's movement. The make up ventilation system that recirculated a part of the cabin air assisted the forward movement of the smoke and fire. Fire hardening of the lavatory compartment using improved materials would result in a higher degree of resistance to ignition and flame spread.

Evolution of Smoke and Toxic Gases

The first smoke visually detected as coming from the lavatory was light in color, possibly from partially wet disposed paper toweling. As it moved forward in the cabin, passengers in the cabin moved forward to vacant seats. However, they were gradually engulfed by the smoke, losing their vision, and were incapacitated. Depressurization of the cabin did not decrease the forward flow of smoke and noxious gases. Visibility was reduced to zero in the aft cabin and smoke entered the flight station filling it, and seriously impairing the pilot's visibility. The pilots, using oxygen masks for breathing, opened the sliding windows in cockpit to restore visibility. Toxic effects, while not accurately known, were severe. After landing, only one passenger was able to escape through the opened galley door.

5.7 Conclusions and Recommendations

The fire safety analysis of this in-flight aircraft fire identifies a number of specific needs and recommendations as follows:

1) The need for a better fire safety technology base, particularly as it relates to materials performance in standard tests and their behavior in actual fire environments.
2) A need for smoke or thermal detection devices in lavatories.
3) The need for fire extinguishment capabilities in lavatories.
4) A need for the fire hardening of lavatories by selection of more fire resistant materials.
5) The need for smoke and toxicity standards for materials employed in aircraft interiors.
6) A need for controlled ventilation of the cabin during an in-flight fire.

Scenario analysis is a powerful, under-utilized tool in our efforts to improve fire safety. Like the "Case Method" in business studies, it permits both the broad or narrow focus of many experts on a defined (perhaps hypothetical) situation. It permits a logical sequence for proceeding from specific to general theorems of behavior in our environment.

Scenarios analysis consistently confirms, as it did in the two preceding examples, that our fire safety development efforts are inadequately supported and, unfortunately, are largely directed to items or components, when, in fact, our problems are those of complex systems.

Polymeric materials are major elements in these complex systems; new materials are being developed and introduced without full understanding of the potential consequences of their use, particularly as they relate to fire safety performance interrelationships. Scenario analysis can assist in better defining the roles that materials can play and what the consequences of their uses may be.